LEED® v3 BUILDING DESIGN & CONSTRUCTION

Exam Study Guide

Sharon J. Sears, Architect, LEED AP

President: Dr. Andrew Temte

Chief Learning Officer: Dr. Tim Smaby

Vice President of Engineering Education: Dr. Jeffrey Manzi, PE

LEED® v3 BUILDING DESIGN & CONSTRUCTION
EXAM STUDY GUIDE

Published by Kaplan Engineering Education

1905 Palace Street

La Crosse, WI 54603

800-420-1432

www.kaplanengineering.com

Printed in the United States of America.

09 10 11 10 9 8 7 6 5 4 3 2 1

ISBN: 1-4277-9244-5
PPN: 1729-1300

CONTENTS

LESSON FOUR

LESSON FIVE

LESSON SIX

ACKNOWLEDGEMENTS

This course was written by Sharon J. Sears. Ms. Sears is a graduate of the School of Architecture (BArch) at the University of Kansas in Lawrence, Kansas and is an architect and LEED®AP BC+C with the Chicago-based architectural firm Interactive Design, Inc.

Consulting editor Kathleen Diane Allen received her undergraduate degree in Economics and Journalism, and her MBA in Finance from the University of Michigan in Ann Arbor, Michigan.

Kaplan would like to especially thank Ronald Fergle, AIA, LEED® AP for his critical review of this book. His feedback and contributions are greatly appreciated.

LESSON ONE 1

INTRODUCTION

INTRODUCTION TO GREEN BUILDING

Green building, also referred to as sustainable building, is the result of a design process focused on the efficient use of resources to reduce the environmental impact of the construction process.

"In the United States, buildings account for:

- 39 percent of total energy use;
- 12 percent of the total water consumption;
- 68 percent of total electrical consumption; and
- 38 percent of the carbon dioxide emissions."[1]

"Green design, construction, and operations have environmental, economic, and social elements that benefit all building stakeholders, including owners, occupants, and the general public."[2]

The Leadership in Energy and Environmental Design (LEED®) Building Certification programs are rating systems designed to rate the sustainability of a construction project. Developed by the United States Green Building Council (USGBC), the first LEED® Pilot Project (LEED® Version 1.0) was launched in 1998. In 2000, Version 2.0 was launched, followed by Version 2.1 in 2002, and Version 2.2 in 2005.

UNITED STATES GREEN BUILDING COUNCIL

Founded in 1993, the USGBC is dedicated to furthering the green movement by providing leadership tools and addressing challenges related to building design, construction, and operation. It is a committee-based, member-driven, and consensus-focused not-for-profit organization, best known for LEED® and Greenbuild. The LEED® rating system is the framework developed for assessing building performance in meeting sustainability goals. Greenbuild is a conference that supports the green building industry focused on environmentally responsible materials, sustainable building design techniques, and public policy.

USGBC also offers many educational opportunities through its Green Building Certification Institute (GBCI), formerly known as the LEED® Accredited Professional (AP) program, which presents education workshops and Web-based seminars to industry professionals and the public.

GREEN BUILDING CERTIFICATION INSTITUTE

The GBCI is a newly incorporated entity established with the support of the USGBC to manage the LEED® AP Exam, oversee the Credentialing Maintenance Program (CMP) for LEED® Accredited Professional (AP), and administer the LEED® building certification program. The USGBC continues to manage development of the LEED® rating systems.

LEED® ACCREDITATION

LEED® 2009 has introduced significant changes in LEED® AP credentialing. There are now three new tiers of LEED® AP accreditation, and a fourth option for existing LEED® APs.

Tier 1—LEED® Green Associate (GA)

The Green Associate (GA) credential is geared toward candidates demonstrating a basic knowledge of green design, construction, and operations. To become a LEED® GA, candidates must complete the following requirements:

- Submit to an application audit
- Agree to the USGBC disciplinary policy
- Submit documented employment in a green or sustainable industry, or enrollment in a relevant educational program
- Submit documented affiliation with LEED® projects
- Successfully complete a core competency examination
- Agree to complete 15 hours of continuing education every 24 months

Tier 2—LEED® AP+

The LEED® AP+ credential is geared toward candidates pursuing advanced knowledge in green building practices. To become a LEED® AP+, candidates must complete the following requirements:

- Submit to an application audit
- Agree to the USGBC disciplinary policy
- Submit documented relevant experience in at least one LEED® project within the last 24 to 36 months
- Successfully complete core competency AND specialization (NC, CI, H, etc.) examinations
- Agree to complete 30 hours of continuing education every 24 months

Tier 3—LEED® AP Fellow

This credential is still under development. It is expected to reward those who have made major contributions to the standards of practice and body of knowledge in the green building field.

Legacy LEED® AP

The new credentialing program allows current LEED® APs to become Legacy LEED® APs; however, to remain an active LEED® AP, all current designees must upgrade to the LEED® AP+ within 24 months of the roll-out of the new credentialing program. To upgrade, designees will need to agree to the USGBC disciplinary policy and continuing education requirements. Designees will be placed directly into one of the Tier II designations based on the exam they originally completed, as well as their general expertise. No additional exam is needed.

LEED® AP BD+C EXAMINATION

The first step in achieving the LEED® AP Building Design and Construction (BD+C) credential is to pass the LEED® Green Associate exam, after which candidates can then take an exam focusing on the LEED® New Construction (NC) rating system. The two exams can be taken together or separately. This study guide focuses on the requirements for the LEED® exam candidate to become a LEED® AP BD+C, which is based on the building certification track for New Construction.

All LEED® credentialing exams are designed to test the candidate's aptitude on the following three hierarchical levels:

- Recognition—Recollection of factual material in context to exam references
- Application—Solving novel problems or scenarios using familiar principles or procedures
- Analysis—The ability to dissect a problem, recognize its different elements, and evaluate the relationships and interaction of its components to be able to create a solution

Exams are administered by computer, but no extensive computer experience is necessary to take the exam. The exam questions are displayed on the screen and the computer records the responses. Test participants may change their answers, skip questions, and flag questions for review.

After being seated at the testing facility, participants are given up to ten minutes to complete a tutorial about how to use the testing software and program. Participants have the option to skip the tutorial and start the test; however, to use as reference during the exam, use this remaining time to note anything about the exam material that may be more difficult or unfamiliar. Participants cannot bring outside items into the exam but will be provided paper and pencil, which must be surrendered upon exiting the exam.

Exam Format:

- Participants have two hours to complete each component of the exam. This does not include the optional ten-minute tutorial before the exam, or the ten-minute survey following the exam.
- The test is comprised of multiple-choice questions, many which have several answers that participants must select. The computer will not allow entry of more than the required number of correct answers.
- The computer will flag any question that does not have the required number of answers selected.
- Participants can mark any question so that they can return to it later.
- At the end of the test, participants will be given a summary of all test questions that

lists which questions are completed, which questions need to be completed, and any questions that the user marked for further review.

- The test questions are presented in a manner that requires participants to dissect the problem. This may require rereading the question and focusing on key words to help identify the intent of the question.
- When the participant has completed the exam, the computer displays the score.

LEED® GREEN ASSOCIATE EXAM

The Green Associate exam tests the candidate's general knowledge of green building practices for both new and existing construction for residential and commercial projects, as well as the skills necessary to support other professionals working on LEED® projects.

The following is the material the Green Associate exam addresses, as identified in the GBCI Handbook:

I. Synergistic Opportunities and LEED® Application Process

   A. Project Requirements (e.g., site; program; budget; schedule)

   B. Costs (e.g., hard costs; soft costs; life-cycle)

   C. Green Resources (e.g., USGBC; Environmental Building News)

   D. Standards that support LEED® Credit (e.g., American Society of Heating; Refrigeration and Air Conditioning D. Engineers [ASHRAE]; Sheet Metal and Air Conditioning Contractors National Association [SMACNA] guidelines; Green Seal)

   E. Credit Interactions (e.g., Energy and IEQ; Waste Management)

   F. Credit Interpretation Rulings/Requests and precedents that lead to exemplary performance credits

   G. Components of LEED® Online and Project Registration

   H. Components of LEED® Score Card

   I. Components of Letter Templates (e.g., project calculations; supplementary documentation)

   J. Strategies to Achieve Credit

   K. Project Boundary; LEED® Boundary; Property Boundary

   L. Prerequisites and/or Minimum Program Requirements for LEED® Certification

   M. Preliminary Rating (target certification level)

   N. Multiple Certifications for Same Building (e.g., Operations and Maintenance for certified building new construction; core and shell and commercial interior; certified building in neighborhood development)

   O. Occupancy Requirements (e.g., for existing buildings, building must be fully occupied for 12 continuous months as described in minimum program requirements)

   P. USGBC Policies (e.g., trademark usage; logo usage)

   Q. Rirements to earn LEED® AP

II. Project Site Factors

   A. Community Connectivity

1. Transportation (e.g., public transportation; bike storage; fuel efficient vehicle parking; parking capacity; car pool parking; car share membership [e.g. Zipcar™]; shuttles; carts)
2. Pedestrian Access (e.g., circulation and accessibility, such as crosswalks, ramps, and trails)

B. Zoning Requirements (e.g., density components, such as calculations—site area and floor area ratio; construction limits; open space; building footprint; development footprint; specific landscaping restrictions)

C. Development Heat Islands (e.g., nonroof; roof; Solar Reflectance Index; emissivity; albedo; heat island effect; green roofs)

III. Water Management

A. Types and Quality of Water (e.g., potable; greywater; blackwater; stormwater)

B. Water Management (e.g., water use reduction through fixtures, such as water closets, urinals, sinks, lavatory faucets, showers; harvesting; baseline water demand; calculations of Full Time Equivalent; irrigation)

IV. Project Systems and Energy Impacts

A. Environmental Concerns (e.g., chlorofluorocarbon [CFC] reduction; no refrigerant option; ozone depletion, fire suppressions without halons or CFCs; phase-out plan; hydrochlorofluorocarbons [HCFC])

B. Green Power (e.g., off-site generated, renewable energy certificates; Green-e providers)

V. Acquisition, Installation, and Management of Project Materials

A. Recycled Materials (e.g., preconsumer; postconsumer; collection requirements; comingled)

B. Locally (regionally) Harvested and Manufactured Materials

C. Construction Waste Management (e.g., written plan; accounted by weight or volume; reduction strategies; polychlorinated biphenyl [PCB] removal and asbestos-containing materials [ACM] management)

VI. Stakeholder Involvement in Innovation

A. Integrated Project Team Criteria (architect; heating, ventilation, and air conditioning [HVAC] engineer; landscape architect; civil engineer; contractor; facility manager)

B. Durability Planning and Management (e.g., material life-cycle; building re-use)

C. Innovative and Regional Design (regional green design and construction measures as appropriate, and established requirements)

VII. Project Surroundings and Public Outreach

A. Codes (e.g., building; plumbing; electrical; mechanical; fire protection)

LEED® BUILDING DESIGN AND CONSTRUCTION EXAM

The BD+C Exam tests the candidate's technical knowledge of the USGBC New Construction Rating System.

The following is the material the BD+C Exam addresses, as identified in the GBCI Handbook:

I. Project Site Factors

A. Considerations for site selection

1. Land issues (e.g., farmland; floodplains [100 year; FEMA maps]; wetlands; water bodies; parkland; brownfield; open space preservation; topography; minimization of disturbed area)

2. Plants and animals (e.g., endangered and imperiled species; native adaptive plants; invasive plant species; Federal wildlife Protection Act; Environmental Protection Agency [EPA] policies; heritage and champion trees; habitat restorations; nesting areas)

B. Community connectivity-services (e.g., nearby amenities; natural amenities; water trails; opportunities for joint use of facilities, such as allowing community access to a school playing field)

C. Development

1. Building and land (e.g., open space; building footprint; development footprint; specific landscaping restrictions, such as use of pollen-free plant materials; visual and/or physical connection to landscape/garden/nature; security requirements that influence site issues such as in healthcare environments [Alzheimer's; psychiatric patients, for example])

2. Lighting (e.g., light pollution; interior and/or exterior light trespass; lighting zone; maximum candela; lighting power densities)

3. Climate conditions (e.g., seasonal changes; solar availability and clear sky data, such as sun path diagrams/charts; day lighting and lighting on streetscapes; precipitation data, such as annual averages; temperature, such as heating and cooling degree days; relative humidity)

II. Water Management

A. Water treatment (e.g., techniques, such as packaged biological removal systems, constructed wetlands, and high-efficiency filtration systems)

B. Stormwater (e.g., rate; imperviousness; pre-development and post-development discharge rate, retention and detention)

C. Irrigation demand (e.g., evapotranspiration; landscape coefficients, such as plant species factor, planting density, microclimate factor)

III. Project Systems and Energy Impacts

A. Energy performance policies (e.g., Minimum Energy Performance; building simulation model; Carbon footprint; intensity-BTU/SF; emissions reductions; building orientation; ASHRAE Advanced Energy Design guides)

B. Building components (e.g., building envelope; HVAC; service water heating; power; lighting; other equipment; lighting power density; receptacle load; insulation; windows)

C. On-site renewable energy (e.g., wind; solar; passive solar; geothermal; biomass; low impact hydro; biogas)

D. Third-party relationship requirements (e.g., prescriptive and performance requirements for Commissioning Agent; design and submittal review; systems manual; follow-up by Commissioning Agent; third-party duty requirements)

E. Energy performance measurement (e.g., measurement verification; building energy simulation vs. metering devices; payment accountability; capability vs. plan; International Performance Measurement and Verification Protocol [IPMVP])

F. Energy tradeoffs (e.g., integration and identification of tradeoffs in energy savings between mechanical, electrical, and building components; lighting design that considers energy use reduction and lighting power density relationship with daylighting)

G. Sources (e.g., central plants; distributed energy [cogeneration]; alternative fuels such as biodiesel, H2 fuel cells, woodchip gasification)

H. Energy usage (e.g., building schedules; occupancy and off-hours; indoor/outdoor air usage rates and impact on energy performance)

IV. Acquisition, Installation, and Management of Project Materials

A. Building reuse (e.g., roof; walls; preplanned conversion of school building to office use)

B. Rapidly renewable materials (e.g., bamboo flooring; cotton batt insulation; wheatboard; cork; wool)

C. Material acquisition (e.g., certified wood; Chain of Custody procedures)

V. Improvements to the Indoor Environment

A. Minimum ventilation requirement (e.g., indoor air quality; natural ventilation; mixed mode ventilation)

B. Tobacco smoke control (e.g., prohibiting smoking; designated smoking area; negative pressure; residential units; weatherstripping; blower door test)

C. Air quality (e.g., carbon dioxide [CO2] concentration; densely occupied spaces vs. non-densely occupied C. spaces; HVAC system integration/automation; air filtration [particulate matter reduction practices])

D. Ventilation effectiveness (e.g., heat recovery strategy; breathing zone; exfiltration/infiltration; breathing zone outdoor airflow, natural ventilation design)

E. Indoor air quality (IAQ)

1. Pre-construction (e.g., development of IAQ Management Program; outdoor air ventilation; ventilation systems design for the reduction in indoor air pollutants)
2. During construction (e.g., protection of absorptive materials from moisture damage; HVAC protection during construction, handling HVAC operation during construction; moisture control, including indoor moisture issues and dehumidification practices; mold prevention and/or remediation)
3. Before occupancy (e.g., HVAC system capability; flush-out procedures; timing requirements for flush-out or testing; air quality testing; contaminant concentrations)
4. During occupancy (e.g., development and implementation of a green

cleaning policy; use of products and materials, equipment, procedures; integrated pest management)

F. Low-emitting materials (e.g., adhesives and sealants; paints; coatings; carpet; composite wood and agrifiber products; VOCs; urea-formaldehyde; VOC budget option; VOC emissions; VOC content)

G. Indoor/outdoor chemical and pollutant control (e.g. entryway systems; walk-off mat contract for cleaning requirement; hazardous gas mixing; pressurize room; door closers; deck-to-deck partitions; MERV 13 for regularly occupied spaces; radon protection; reduction practices for radon gas; other soil gas contaminants)

H. Lighting controls (e.g., individual occupant control; task lighting; dual levels; occupancy sensors; daylight sensors; building automation system; window and floor area; geometry factor; visible transmittance; shading devices; light shelves; skylights)

I. Thermal controls (e.g., individual occupant control of thermal comfort; operable windows; thermostat; diffusers; radiant panel options; building automation system; thermal comfort design; written plan for corrective action; distribution of space heating and cooling)

J. Views (e.g., through interior glazing, vision glazing; glare effects on individuals, such as patients or students)

K. Types of building spaces (e.g., regularly occupied spaces; individual occupant work spaces; group multi-occupant spaces; classroom and core learning spaces)

VI. Stakeholder Involvement in Innovation

A. Design workshop/charrette (meeting on integrated green strategies)

B. Ways to earn credit (e.g., innovative methods; building as a teaching tool; tailor lighting color to task; green educational program; residential construction methodologies; tenant guidelines; documentation of sustainable building cost impacts)

C. Education of building manager (development of a building manual and a demonstration walkthrough of the green features in the building)

VII. Project Surroundings and Public Outreach

A. Infrastructure (e.g., access information to sewer and water supply service areas; municipal utilities, such as availability and capacity of existing lines, future demand, power and water management district)

B. Zoning requirements (e.g., land use change amendments; public hearings)

C. Government planning agencies (e.g., Environmental Protection Agency [EPA]; local; state; federal; USDA; Public Health Code)

D. Reduced parking methods (e.g., shared parking facilities; carpools; car-share; bike secure parking)

E. Transit oriented development (e.g., access to train; bus; multi-modal interfaces)

F. Pedestrian-oriented streetscape design (bike and pedestrian connectivity; vehicular traffic interface; bike lanes)

LEED® RATING SYSTEMS

The USGBC currently offers nine LEED® rating systems. Each rating system addresses standard sustainability issues, as well as sustainability issues specific to each building type. They are:

1. New Construction—NC
2. Existing Buildings: Operations and Maintenance—EB
3. Commercial Interiors—CI
4. Core and Shell
5. Schools
6. Retail
7. Health Care
8. Homes
9. Neighborhood Development (currently a pilot program)

The LEED® NC Rating System assesses a building's sustainability based on accepted energy and environmental principles.

The rating system is divided into seven topics:

1. Sustainable Sites
2. Water Efficiency
3. Energy and Atmosphere
4. Materials and Resources
5. Indoor Environmental Quality
6. Innovation in Design
7. Regional Priority

The system includes prerequisites and credits. A prerequisite is a mandatory requirement; a project cannot attain LEED® certification without fulfilling all of the prerequisites in the LEED® Rating System. A credit is an individual requirement that teams can earn one or more points by satisfying. The points a team earns are added together to determine if the project will receive a certified, silver, gold or platinum certification. The more points earned, the higher the certification level.

There are 100 base points in the LEED® NC Rating System, plus six possible Innovation in Design points, and four Regional Priority points. The new thresholds for LEED® NC certification levels are as follows:

- **Certified**: 40–49 points
- **Silver**: 50–59 points
- **Gold**: 60–79 points
- **Platinum**: 80 points or more

INNOVATION IN DESIGN

Innovation in Design is the opportunity for design teams to earn points for using new technologies introduced to the marketplace, or for exceptional performance above the requirements set by the LEED® Green Building Rating System. An additional point can be earned for having a LEED® AP as an integral member of the team.

Teams can earn Innovation in Design points for exemplary performance by greatly exceeding the performance level, or by expanding the scope of a credit. To attain the exemplary performance credit, teams must first meet the credit criteria, and then achieve the next performance threshold. If a credit has more than one compliance path, the team can also achieve an Innovation in Design point if it satisfies more than one compliance path.

REGIONAL PRIORITY

New to LEED® 2009 is the potential for teams to earn Regional Priority points. To encourage design teams to focus on region-specific environmental issues, USGBC Regional Councils have identified environmental zones, based on ZIP codes, and allocated six credits within each zone eligible to receive regional priority points. Information regarding the zones, separated by state and then ZIP code, can be downloaded at *www.usgbc.org*.

Projects qualifying for Regional Priority will earn one point for meeting the Regional Priority criteria, and one point for meeting the credit. Projects can earn up to four Regional Priority credits.

CREDIT WEIGHTINGS

Also new to LEED® 2009 is the addition of Credit Weightings, which have been added to respond to new expectations that green buildings can also affect social, economic and environmental issues, as well as changes in market conditions and user requirements. New credit weightings are focused on energy use, water use, materials, solid waste, and transportation-related credit criteria. Credits with stronger energy, water, and transportation impacts are eligible for more points under the new weighting system.

Credit values that have changed:

- A total of 100 points are possible, excluding Innovation in Design and Regional credits
- Five Innovation in Design credits are now available
- Four Regional Priority credits are now available
- Allocation of points among credit categories has changed, resulting in a change in the relative emphasis of the categories

Credit values that remain unchanged:

- Existing credits have not changed
- Each credit is worth a minimum score of one
- Credits are positive whole numbers, with no fractions or negative values
- Credits have a consistent range of point values, regardless of location or connections among credits

To arrive at the credit weights, many factors were taken into account. LEED® 2009 Credit Weightings are documented in a self-contained Microsoft Excel workbook, which includes all calculations and rules used to assign weights to individual LEED® credits. The calculations in the workbook were developed using the U.S. Environmental Protection Agency (EPA) Tool for Reduction and Assessment of Chemical and other Environmental Impacts (TRACI), using the environmental categories as the basis for weighting each credit. The National Institutes of Standard and Technology (NIST) was also used to compare the impact categories and to add a relative weight to each credit in regard to the importance of the building-related impacts.

CREDIT INTERPRETATION REQUESTS

Credit Interpretation Requests (CIRs) and the ruling process allows teams to gain technical and administrative direction regarding how Minimum Project Requirements, prerequisites, and credits relate to their projects. All LEED® 2009 CIRs are project-specific. They can be submitted after the project is registered, and expire at the time of the final award or denial of project certification. CIRs can only apply to one LEED® Requirement and are not transferable to other LEED® projects.

Teams submit CIRs using LEED®-Online, using either a form for a particular prerequisite or credit or, if the team's question is specific to a rating system requirement, teams can use a stand-alone CIR form. Teams must submit CIRs and their rulings with their LEED® application to ensure all necessary components are considered at the time of review.

When submitting a CIR form, teams must follow specific instructions, including:

- not submitting a CIR in the form of a letter;
- providing only essential background information in the CIR;
- verifying that the CIR does not exceed 600 words or 4,000 characters, including spaces; and
- ensuring that the CIR does not include attachments, cut-sheets, plans, or drawings.

The CIR process cannot change prerequisite or credit language, nor guarantee that LEED® Minimum Project Requirements, prerequisites, and credits are satisfied. Teams pursuing LEED® 2009 certification cannot reference CIRs from pre-LEED® 2009 rating systems, and project teams must adhere to the CIR rulings received for their project. The project team is still required to demonstrate that all credit requirements are satisfied.

MINIMUM PROGRAM REQUIREMENTS

Minimum Program Requirements (MPRs) have also been introduced in LEED® 2009 as specific requirements that projects seeking LEED® certification must follow. The intent is to provide clear guidance to consumers, protect the integrity of the LEED® program, and reduce complications that occur during the LEED® certification process.

The following guidelines are only applicable to projects certifying under LEED® 2009, including projects that migrate to LEED® 2009 from earlier versions of LEED®. Only MPRs in place at the time a project registers or upgrades will apply to the project. MPRs will be periodically updated and posted on the GBCI Web site in a document titled *LEED® 2009 MPR Supplement Guidance*, which can be accessed at *www.GBCI.org*.

A project's certification can be revoked and its registration and/or certification fees will not be refunded if it is determined that a project does not meet any applicable MPRs.

Currently established MPRs for LEED® NC include the following:

- "1. MUST COMPLY WITH ENVIRONMENTAL LAWS

  The LEED® project building or space, all other real property within the LEED® project boundary, and all project work LEED® project registration must comply with all applicable federal, state, and local building-related environmental laws and regulations in place where the project is located. This condition must be satisfied from the date of, or the initiation of, schematic design, whichever comes first, until the date that the building receives a certificate of occupancy or similar official indication that it is ready for use.

- 2. MUST BE A COMPLETE, PERMANENT BUILDING OR SPACE

  All Rating Systems:

  All LEED® projects must be designed for, constructed on, and operated on a permanent location on existing land. No building or space that is designed to move at any point in its lifetime may pursue LEED® certification.

LEED® projects must include the new, from the ground-up design and construction, or major renovation, of at least one building in its entirety.

Additionally, construction prerequisites and credits may not be submitted for review until **substantial completion of construction** has occurred.

- 3. MUST USE A REASONABLE SITE BOUNDARY

New Construction, Core and Shell, Schools, Existing Buildings: Operations and Maintenance

1. The LEED® project boundary must include all contiguous land that is associated with and supports normal building operations for the LEED® project building, including all land that was, or will be disturbed for, the purpose of undertaking the LEED® project.
2. The LEED® project boundary may not include land that is owned by a party other than that which owns the LEED® project, unless that land is associated with and supports normal building operations for the LEED® project building.
3. LEED® projects located on a campus must have project boundaries such that if all the buildings on campus become LEED® certified, then 100 percent of the gross land area on the campus would be included within a LEED® boundary. If this requirement is in conflict with MPR #7: Must Comply with Minimum Building Area to Site Area Ratio, then MPR #7 takes precedence.
4. Any given parcel of real property may only be attributed to a single LEED® project building.
5. Gerrymandering of a LEED® project boundary is prohibited: the boundary may not unreasonably exclude sections of land to create boundaries in unreasonable shapes for the sole purpose of complying with prerequisites or credits.

- 4. MUST COMPLY WITH MINIMUM FLOOR AREA REQUIREMENTS

New Construction, Core and Shell, Schools, Existing Buildings: Operations and Maintenance

The LEED® project must include a minimum of 1,000 square feet (93 square meters) of gross floor area.

- 5. MUST COMPLY WITH MINIMUM OCCUPANCY RATES

New Construction, Core and Shell, Schools, and Commercial Interiors:

Full Time Equivalent Occupancy

The LEED® project must serve one or more Full Time Equivalent (FTE) occupant(s), calculated as an annual average in order to use LEED® in its entirety. If the project serves less than one annualized FTE, optional credits from the Indoor Environmental Quality category may not be earned (the prerequisites must still be earned).

- 6. MUST COMMIT TO SHARING WHOLE-BUILDING ENERGY AND WATER-USAGE DATA

All certified projects must commit to sharing with USGBC and/or GBCI all available actual whole-project energy and water-usage data for a period of at least 5 years. This period starts on the date that the LEED® project begins typical physical occupancy if certifying under New Construction, Core and Shell, Schools, or Commercial Interiors, or the date that the building is awarded certification if certifying under Existing Buildings: Operations & Maintenance. Sharing this data includes supplying information on a regular basis in a free, accessible, and secure online tool or, if necessary, taking any action to authorize the collection of information directly from service or utility providers. This commitment must carry forward if the building or space changes ownership or lessee.

- 7. MUST COMPLY WITH A MINIMUM BUILDING-AREA TO SITE-AREA RATIO

The gross floor area of the LEED® project building must be no less than 2 percent of the gross land area within the LEED® project boundary."[3]

LEED®—ONLINE

LEED®—Online was introduced with LEED® NC 2.2. It is a tool designed to allow members of a project team to manage the LEED® documentation process, and to verify compliance with the LEED® rating system. It allows project teams to manage project details, complete documentation requirements for LEED® credits and prerequisites, upload supporting files, submit applications for review, receive reviewer feedback, and earn LEED® certification online.

LEED®—Online Version 3 is available only for those projects registered under LEED® 2009 and offers several enhancements over its predecessor. Version 3 provides a more intuitive user interface, enhances communication between project teams and certifying bodies, and contains upgrades that respond to the changes in the LEED® 2009 rating system.

ABBREVIATIONS AND ACRONYMS

CI	Commercial Interiors
EB	Existing Buildings
GBCI	Green Building Certification Institute
H	Homes
LEED®	Leadership in Energy and Environmental Design
LEED® AP	LEED® Accredited Professional
LEED® AP+	LEED® Accredited Professional Plus
LEED® BD+C	LEED® Building Design and Construction
LEED® GA	LEED® Green Associate
MPR	Minimum Program Requirements
NC	New Construction
U.S. EPA	United States Environmental Protection Agency
USGBC	United States Green Building Council

FOOTNOTES

1. U.S. Environmental Protection Agency, *Why Build Green?*, 2009, http://www.epa.gov/greenbuilding/pubs/whybuild.htm.
2. Arizona State University, *Green Building Design—Why Design Green?,* 2009, http://uabf.asu.edu/green_building_design.
3. U.S. Green Building Council, *LEED® 2009 for New Construction and Major Renovation*, 2009, http://www.leedbuilding.org/ShowFile.aspx?DocumentID=5546.

LESSON 1 QUIZ

1. Which three of the following credits are eligible to attain an Innovation in Design point for exemplary performance?

   A. SSc5 Site Development

   B. SSc8 Light Pollution Reduction

   C. WEc2 Innovative Wastewater Technologies

   D. EAc1 Optimized Energy Performance

   E. EAc4 Enhanced Refrigeration Management

   F. IEQc2 Increased Ventilation

2. Upon award of LEED® Certification to a project, the USGBC will (choose one)

   A. organize the building dedication.

   B. publish the project's achievements via news releases, Web sites, etc.

   C. provide marketing.

   D. organize open house tours of the project.

   E. help find funding for local community green outreach.

3. Which three of the following can become a USGBC National Member?

   A. Students

   B. State, local, and federal agencies

   C. Universities

   D. Real estate agencies

   E. Interns

   F. Employees of an architectural firm

4. In what year was the USGBC founded?

   A. 1999

   B. 1993

   C. 2000

   D. 2003

5. What are the three fundamental organizational characteristics of the USGBC?

   A. Committee-based

   B. Individual

   C. Member-driven

   D. Consensus-focused

   E. Profit-focused

6. Beginning with LEED® 2009, test candidates who demonstrate a basic knowledge of design, construction, and operations can become which one of the following?

   A. LEED® Accredited Professional

   B. LEED® Green Associate

   C. LEED® Green Partner

   D. LEED® Certified

   E. LEED® Apprentice

7. What are the new LEED® NC thresholds for achieving LEED® certification?

   A. 40–59 Certified
60–79 Silver
80–99 Gold
100 or more Platinum

   B. 40–49 Certified
50–59 Silver
60–79 Gold
80 or more Platinum

   C. 26–32 Certified
33–38 Silver
39–51 Gold
52–69 Platinum

   D. 26–35 Certified
36–42 Silver
43–51 Gold
52–69 Platinum

8. Which three of the following are new to LEED® 2009?

A. Credit Weightings

B. Credit Interpretation Requests

C. Regional Priority

D. LEED®—Online

E. Minimum Program Requirements

F. Exemplary Performance

9. A shirt manufacturer is planning to expand their production capabilities. The manufacturer has purchased a vacant building that is of sufficient size to house their expanded operations. Modifications to the building will include a new HVAC system, the façade will receive a new curtain wall system, and the building's interior will receive new partitions and finishes. Which LEED® rating system should the manufacturer use to attain a LEED® certification?

A. Existing Buildings

B. Core and Shell

C. Commercial Interiors

D. New Construction

10. Credit Weightings are new to the LEED® rating system with the introduction of LEED® 2009. This category was introduced to respond to social, economic, and environmental issues, changes in market conditions, and user requirements. The new credits are focused on which three of the following?

A. Energy use

B. Air quality

C. Water use

D. Transportation

E. Refrigeration

F. Recycling

QUIZ ANSWERS

Lesson 1

1. **A, C, D** For a complete list of credits that can attain an Innovation in Design point for exemplary performance, refer to Lesson 7, Innovation in Design. No IEQ credits are eligible for exemplary performance credits.
2. **B** USGBC typically does not visit the project, or provide funding sources or other local promotion.
3. **B, C, D** Only organizations can belong to USGBC National; individuals can belong to local USGBC chapters.
4. **B** The USGBC was founded in 1993.
5. **A, C, D** The USGBC is a committee-based, member-driven, and consensus-focused not-for-profit organization dedicated to furthering the green movement by providing leadership tools, and responding to green building challenges in how buildings are designed, built, and operated.
6. **B** Beginning with LEED® 2009, test candidates can become a Green Associate by demonstrating a basic knowledge of sustainable design. Candidates can become a LEED® Accredited Professional by taking a second test focused on the technical aspects of one of the following: New Construction, Commercial Interiors or Existing Buildings: Operations and Maintenance.
7. **B** New with LEED® 2009, there are 100 base points, plus six possible points for Innovation in Design, and four possible points for Regional Priority, for a potential total of 110 points.
8. **A, C, E** Credit Weightings, Regional Priority, and Minimum Program Requirements were added in LEED® 2009 to address concerns regarding application of the rating system across different regions and strength of environmental impact of credits.
9. **D** The New Construction Rating System is the best choice for this project since it involves extensive facade work, new mechanical systems, and new interior partitions and finishes.
10. **A, C, D** Credits with stronger energy, water, and transportation impacts are eligible for more points under the new weighting system.

LESSON TWO

SUSTAINABLE SITES

INTRODUCTION

The earliest decisions made on a project are typically those regarding selection of the site and how it will be developed. These decisions include density, size of building footprint size, stormwater management, brownfield redevelopment, green roof, building orientation, materials used to develop site amenities, proximity to transportation networks, and site lighting.

When establishing goals for the project, each of these issues is important to consider. Decisions made during site selection can impact other LEED® credits and how the building and site will use natural and man-made resources.

In this chapter, we will look at the prerequisite and credits for the **Sustainable Sites** category, summarized below:

SSp1 Construction Activity Pollution Prevention

SSc1 Site Selection .... 1 Point

SSc2 Development Density and Community Connectivity .... 5 Points

SSc3 Brownfield Redevelopment .... 1 Point

SSc4.1 Alternative Transportation—Public Transportation Access .... 6 Points

SSc4.2 Alternative Transportation—Bicycle Storage and Changing Rooms .... 1 Point

SSc4.3 Alternative Transportation—Low-Emitting and Fuel-Efficient Vehicles .... 3 Points

SSc4.4 Alternative Transportation—Parking Capacity .... 2 Points

SSc5.1 Site Development—Protect or Restore Habitat .... 1 Point

SSc5.2 Site Development—Maximize Open Space .... 1 Point

SSc6.1 Stormwater Design—Quantity Control .... 1 Point

SSc6.2 Stormwater Design—Quality Control .... 1 Point

SSc7.1 Heat Island Effect—Nonroof..... 1 Point

SSc7.2 Heat Island Effect—Roof .... 1 Point

SSc8 Light Pollution Reduction............ 1 Point

Evaluation of potential sites should include sustainable design objectives. Ecologically sensitive resources, such as prime farmland, parks, and wetlands should be avoided because these types of sites provide habitats for plants, wildlife, and livestock. To be able to take advantage of existing infrastructure, consider the proximity to transportation and utility networks. Construction activity can disrupt local and regional ecosystems in a variety of ways. When topsoil is removed or damaged, erosion carries soil to receiving streams and other waterways, reducing water quality and increasing pollution. Understanding how early decisions can minimize or reduce the impact of new construction is important when developing a sustainable site.

Reducing the extent of site development can (1) minimize the impact of new construction on ecosystems; (2) decrease strain on existing drainage systems by allowing stormwater runoff to be absorbed into the site in lieu of immediate discharge into the city water system; and (3) limit the environmental impact of building on local ecosystems. Orienting the building to take advantage of solar angles helps to minimize heating and cooling loads. Selecting an existing urban site helps to preserve natural resources and allows projects to take advantage of existing urban infrastructure, including utilities, roads, bike paths, and mass transportation networks. Other site developments that assist these goals include reducing impervious paving, and/or using pervious pavers. Using low-albedo site paving and roofing materials reduces heat island effect. Green roofs also reduce heat island effect, and decrease stormwater runoff by absorbing stormwater into the soil of the green roofs and filtering it before discharging into the stormwater sewer system. Using light fixtures with full cutoffs minimizes light trespass in order to increase night sky visibility and reduce development impact on nocturnal environments.

SSp1 CONSTRUCTION ACTIVITY POLLUTION PREVENTION

Intent

"By most accounts, the most environmentally dangerous period of development is the initial construction phase when land is cleared of vegetation and graded to create a proper surface for construction. The removal of natural vegetation and topsoil makes the exposed area particularly susceptible to erosion, causing transformation of existing drainage areas and disturbance of sensitive areas."[1] The intent of the Construction Activity Pollution Prevention prerequisite is to control erosion, sedimentation, and dust generated during construction. **Erosion** refers to the wearing away of soil through natural action. **Sedimentation** is the addition of soil to bodies of water through natural and human activities. Sedimentation decreases the quality of water and accelerates the aging process of these bodies of water.

Requirements

To achieve this prerequisite, the project team must develop and implement an Erosion and Sedimentation Control (**ESC**) plan for all construction activities. The plan must comply with either the 2003 EPA Construction General Permit (**CGP**) or local erosion and sediment control standards, whichever is more stringent.

The 2003 EPA CGP includes phases I and II of the National Pollutant Discharge Elimination System (**NPDES**). The requirements of this standard apply only to projects greater than one acre in size, but LEED® NC requires the standard to be applied to all projects seeking LEED® certification.

Implementation of soil and sedimentation control measures minimizes the off-site consequences of erosion from developed sites. Runoff from developed sites carries pollutants, sediments, and excess nutrients that can cause extensive damage to aquatic habitats in the receiving waters. Runoff can include nitrogen and phosphorous, which can cause unwanted plant growth in aquatic systems and reduce the population diversity of indigenous fish, plants, and animals.

Sedimentation can add to the degradation of water bodies and aquatic habitats. Sedimentation buildup can restrict water flow in streams, increasing flooding and turbidity. Turbidity reduces sunlight penetration into water and can lead to reduced photosynthesis in aquatic vegetation, resulting in lower oxygen levels that cannot support diverse communities of aquatic life.

Dust generation from construction activity can cause both health and environmental impacts. Breathing dust can be linked to asthma, decreased lung function, and other breathing issues. Dust can also be blown for long distances before settling in lakes and streams, increasing their acidity and changing their nutrient balance.

Referenced Standard

The referenced standard for this prerequisite is *Storm Water Management for Construction Activities*, Chapter 3 (U.S. EPA No. EPA 832R92005). This standard can be downloaded at *www.epa.gov/npdes/pubs/chap03_conguide.pdf*. This standard describes stabilization and structural control, the two methods used to control erosion and sedimentation. **Stabilization** includes temporary seeding, permanent seeding, and mulching. **Structural control** includes silt fences, earth dikes, sediment traps, and sediment basins. The type of control method used depends on specific site conditions. Depending on the jurisdiction of a project's site, NPDES phases I and II, which could include Stormwater Pollution Prevention Plan (**SWPPP**), may already be required. The framework for NPDES phases I and II and SWPPP are included in the referenced standard for this credit; only verification of compliance with the prerequisite and implementation of the ESC plan would be required for the prerequisite. If the local jurisdiction does not require an ESC plan, the referenced standard should be used as the basis for the project's ESC plan.

The ESC plan should be developed during the design phase of the project and incorporated into the construction documents. It must address the following issues:

- Preventing soil loss during construction caused by stormwater runoff and/or wind erosion, including protecting topsoil by stockpiling it for reuse
- Preventing sedimentation of storm sewer or streams
- Preventing air pollution caused by the generation of dust

The ESC plan should include a statement describing erosion and stormwater control objectives, a comparison of pre- and post-development stormwater runoff conditions, and a description of maintenance required for the implemented plan.

Related Credits

Minimizing site disturbance and restoring natural habitats can also help teams achieve SSc5.1 Site Development—Protect or Restore Habitat and SSc5.2 Site Development—Maximize Open Space.

Teams choosing to limit site disruption to preserve the natural hydrology of the site may also be able to achieve SSc6.1 Stormwater Design—Quantity Control and SSc6.2 Stormwater Design—Quality Control.

SSc1 SITE SELECTION

Intent

The decisions made by the team during this time in the project's development can affect many other aspects and costs of the project. Choosing an undeveloped site could result in the need to bring roads and utilities into undeveloped areas. By giving careful consideration to the guidelines for this credit, teams can avoid selecting an inappropriate site to develop for a new construction project.

Requirements

To achieve this credit, the project team cannot develop buildings, roads, or parking areas on portions of sites that meet any of the following criteria:

- Prime farmland
- Previously undeveloped land where the lowest site elevation is lower than five feet above the current 100-year flood level as defined by Federal Emergency Management Agency (FEMA)
- Home to any species on a federal or state threatened or endangered list
- Within 100 feet of any wetlands
- Public park land
- Previously undeveloped land within 50 feet of a water body that could support fish, recreation or industrial use, consistent with the Clean Water Act

During the site selection process, teams should identify previously developed sites that fit the needs of the project. Once a site has been selected, teams should identify the site's natural features and the measures needed to preserve them. Some of the site's existing features, such as natural shelter from trees or terrain, natural areas for outdoor activities, and water features can be incorporated into the design of the site.

Development of a greenfield site will impact existing wildlife and natural habitats and will require new infrastructure to be brought to the site. Teams developing greenfield sites also run the risk of losing public support by developing a site that is inappropriate. There is also the possibility the project will suffer property damage, such as floods, landslides, and sinkholes as a result of selection of an inappropriate site.

The decision to construct on a previously developed site allows teams to take advantage of the economics of existing urban infrastructure, including roads, mass transit systems and utilities, and preserves non-urban areas for wildlife, recreation, and ecological balance.

Referenced Standards

There are five referenced standards supporting this credit. Each standard addresses a specific area of conservation addressed by this credit.

1. U.S. Department of Agriculture Definition of Prime Agricultural Lands as stated in United States Code of Federal Regulations, Title 7, Volume 6, Parts 400–699, Section 65705 (Citation 7CFR657.5). The standard reads: "Prime farmland is land that has the best combination of physical and chemical characteristics for producing food, feed, forage, fiber, and oilseed crops, and is also available for these uses (the land could be cropland, pastureland, rangeland, forestland, or other land, but not urban built-up land or water). It has the soil quality, growing season, and moisture supply needed to economically produce sustained high yields of crops when treated and managed, including water management, according to acceptable farming methods. In general, prime farmlands have an adequate and dependable water supply from precipitation or irrigation, a favorable temperature and growing season,

acceptable acidity or alkalinity, acceptable salt and sodium content, and few or no rocks. They are permeable to water and air. Prime farmlands are not excessively erodible or saturated with water for a long period of time, and they either do not flood frequently or are protected from flooding. Examples of soils that qualify as prime farmland are Palouse silt loan, 0 percent to 7 percent slopes; Brookston silty clay loam, drained; and Tama silty clay loam, 0 percent to 5 percent slopes."

2. FEMA 100-Year Flood Definition. This definition states: "The term '100-year flood' is misleading. It is not the flood that will occur once every 100 years. Rather, it is the flood elevation that has a 1 percent chance of being equaled or exceeded each year. Thus, the 100-year flood could occur more than once in a relatively short period of time. The 100-year flood, which is the standard used by most Federal and state agencies, is used by the NFIP as the standard for floodplain management and to determine the need for flood insurance. A structure located within a special flood hazard area shown on an NFIP map has a 26 percent chance of suffering flood damage during the term of a 30 year mortgage."
3. Endangered Species Lists (U.S. Fish and Wildlife Service, List of Threatened and Endangered Species). This list can be downloaded at *www.fema.gov*. The standard describes threatened and endangered wildlife and plants and provides a current list of the country's native plants and animals that are candidates for addition to the list.
4. National Marine Fisheries Services, List of Endangered Marine Species. Download this list at *http://nmfs.noaa.gov/pr/species/esa_species.htm*. Also refer to your state agencies for state-specific lists of endangered or threatened wildlife and plant species.
5. Definition of Wetlands in the United States Code of Federal Regulations, 40 CFR, Parts 230–233, and Part 22. This definition can be downloaded at *www.gpoaccess.gov/cfr/index.html*. This standard addresses wetlands and dredged discharge or filled materials into state regulated waters. The definition of wetland as it pertains to this section is found in Part 230 and states: "Wetlands consist of areas that are inundated or saturated by surface or groundwater at a frequency and duration sufficient to support, and that under normal circumstances do support, a prevalence of vegetation typically adapted for life in saturated soil conditions."

Related Credits

Access to public transportation and community services is likely to be available to projects located on previously developed sites. These sites are also likely to require remediation. In addition, choosing to minimize the building footprint, increase and/or protect open space can help teams meet the requirements for the following credits:

- SSc2 Development Density and Community Connectivity
- SSc3 Brownfield Redevelopment
- SSc4.1 Alternative Transportation—Public Transportation Access
- SSc5.1 Site Development—Protect or Restore Habitat
- SSc5.2 Site Development—Maximize Open Space
- SSc6.1 Stormwater Design—Quantity Control
- SSc6.2 Stormwater Design—Quality Control

SSc2 DEVELOPMENT DENSITY AND COMMUNITY CONNECTIVITY

Intent

Guiding development toward previously developed urban sites with pedestrian access to community services helps the teams use existing infrastructure, protect greenfields, and preserve habitat and natural resources. Selecting an urban site allows the building's end users to reduce their reliance on automobiles, which helps to restore and strengthen established urban living patterns and support stable and interactive communities.

"Growth and development can give a community the resources needed to revitalize a downtown, refurbish a main street, build new schools, and create vibrant places to live, work, shop and play. The environmental impacts of development, however, can pose challenges for communities striving to protect their natural resources. Development that uses land efficiently and protects undisturbed natural lands allows a community to grow and still protect its water resources."[2]

Requirements

Teams have two options to achieve this credit. Option 1 is to achieve Development Density by constructing or renovating a building on a previously developed site within a community with a minimum density of 60,000 square feet per acre. Option 2 is to achieve Community Connectivity by constructing or developing a site that is within ½ mile of a neighborhood or residential zone that has an average density of ten units per acre net, is within ½ mile of at least ten basic services, and pedestrian access is provided between the building and the services. Basic services include, but are not limited to:

- Bank
- Beauty
- Cleaners
- Community Center
- Convenience Grocery
- Day Care
- Fire Station
- Fitness Center
- Hardware
- Laundry
- Library
- Medical/Dental
- Museum
- Park
- Pharmacy
- Place of Worship
- Post Office Facility
- Restaurant
- School
- Senior Care Facility
- Supermarket
- Theatre

To determine development density, the density of the site and surrounding sites need to be determined. First, determine the square footage of the project site and the surrounding sites. If the site is part of a larger site, such as a campus, the project area must be defined and used consistently throughout the project. Development density is calculated using the following equation. Note that density must be at least 60,000 square foot per acre to qualify for this credit.

$$\text{Development Density (sf/acre)} = \frac{\text{Gross Building Area (sf)}}{\text{Site Area (acres)}}$$

To calculate the density radius, convert the site area into square feet and calculate the square root of the result. This normalizes the calculation by removing the effects of the site shape.

Density Radius (lf) =

$$3 \times \sqrt{[\text{Site (acres)} \times 43{,}560 \text{ (sf/acre)}]}$$

To determine the average property density within the density boundary, add the square footages and site areas for each property and divide the total square feet by the site area. The average property density of the property within the boundaries must be at least 60,000 square feet per acre.

Average Property Density within Density

$$\text{Boundary} = \frac{\sum \text{Square Footage}}{\sum \text{Site Area}}$$

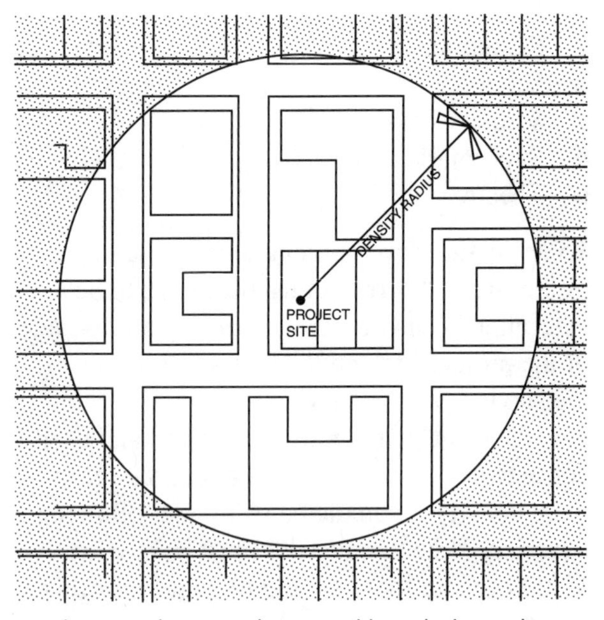

Development density is determined by calculating the building space and site area for each property located within the density radius. These values are totaled, and the average density is calculated by dividing the total building square footage by the total site area.

Figure 2.1 Density Radius Sample Map

To determine community connectivity, both residential and commercial neighbors should be considered. On a site map, draw a ½-mile radius beginning at the main entrance to the building and identify all basic services and residential developments with ten or more units per acre within the radius.

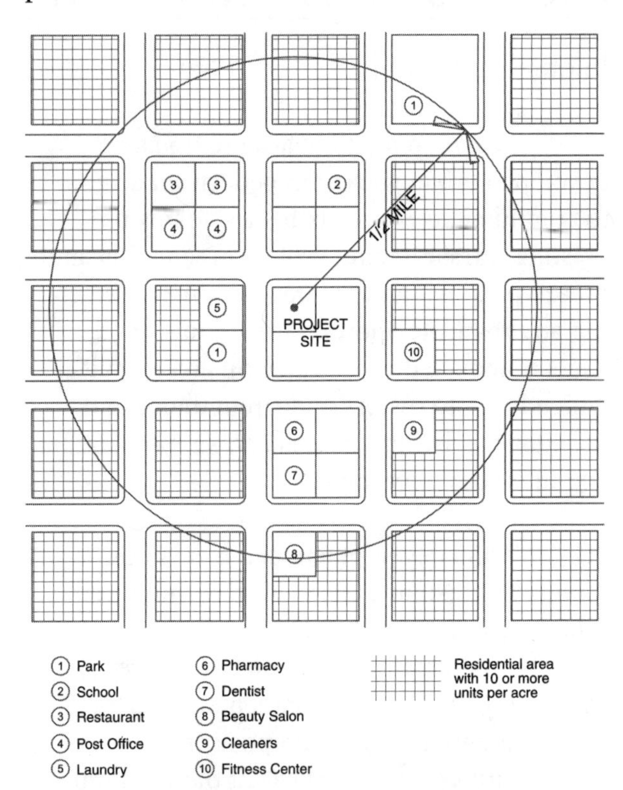

Figure 2.2 Community Connectivity Sample Map

Mixed-use projects can count one service within the project boundary toward the required ten basic services. That service must be open to the public. At least eight of the basic services must be existing and operational at the time of the building's occupancy. No more than two of the basic services can be anticipated, and those must be operational within one year of the building's occupancy.

Exemplary Performance

One Innovation in Design point for exemplary performance is available to teams that, after meeting the requirements for Option 1, meet one of the following:

1. The density for the project itself is at least double that of the average density within the calculated area.

   OR

2. After doubling the density radius determined in the base credit, the density within the larger radius is at least 120,000 square feet per acre.

Related Credits

Projects located in urban areas will probably be located on previously developed land that is near public transportation, which can help teams meet the requirements for SSc1 Site Selection and SSc4.1Alternative Transportation—Public Transportation Access.

SSc3 BROWNFIELD REDEVELOPMENT

Intent

When brownfields sit unused, everybody loses. Neighbors face environmental worries and reduced property values. Cities see roads, sewers, and other infrastructure underused. When owners or developers clean up brownfields and put them to new uses, many people benefit. Environmental remediation addresses environmental problems. Redevelopment can bring new jobs and higher tax revenues to communities.

Development of a brownfield site can help a team gain economic and community-related benefits. Often, brownfield sites are situated in desirable locations, are less expensive, and are located near public transportation and community services.

In the past, developers have been reluctant to develop brownfield sites because of the cost and potential liability associated with the existing contamination. The EPA and many state and local agencies now provide incentives for brownfield development by reducing liability for those who choose to remediate these sites. Tax Increment Financing (TIFs) offered by local governments, as well as other incentives, may be available to teams choosing to remediate a brownfield site.

Requirements

To achieve this credit, teams need to develop a site that is contaminated or classified as a brownfield as defined by ASTM E1903-97 Phase II Environmental Site Assessment or identified as a brownfield by a local, state, or federal government agency.

The remediation is site-specific, but proven technologies that do not damage above ground or underground natural features should be implemented. If contaminated groundwater is found, it is typically pumped to the surface, removed from the site and treated with either a chemical or physical process. If contaminated soils are encountered, depending on the type of contamination, the soil can be treated in place using a process called in situ remediation, or the soil can be removed from the site and taken to an appropriate disposal site.

Referenced Standards

There are three referenced standards for this credit as listed below.

1. U.S. EPA, Definition of Brownfields. This definition can be downloaded at *www.epa.gov/brownfields/glossary.htm* and states: "With certain legal exclusions and additions, the term 'brownfield site' means real property, the expansion, redevelopment, or reuse of which may be complicated by the presence or potential presence of a hazardous substance, pollutant, or contaminant." (Public Law 107-118 (H.R. 2869)—"Small Business Liability Relief and Brownfields

Revitalization Act" signed into law January 11, 2002.)

2. ASTM E1527-05, Phase I Environmental Site Assessment (ESA). This assessment can be downloaded at *www.astm.org*. A Phase I ESA is a report prepared for a real estate property, which identifies potential or existing environmental contamination liabilities. The report typically addresses both the underlying land, as well as physical improvements to the property; however, techniques applied in a Phase I ESA never include actual collection of physical samples or chemical analyses of any kind. Scrutiny of the land includes examination of potential soil contamination, groundwater quality, surface water quality, and sometimes issues related to hazardous substance uptake by biota. The examination of a site may include one or more of the following actions: definition of any chemical residues within structures; identification of possible asbestos-containing building materials; inventory of hazardous substances stored or used on site; assessment of mold and mildew, and evaluation of other indoor air quality parameters.
3. ASTM E1903-97, Phase II Environmental Site Assessment. This assessment can be downloaded at *www.astm.org*. If needed, a Phase II ESA is typically recommended in the Phase I ESA, but can be implemented without a Phase I ESA if the interested parties are aware of existing contamination. The Phase II ESA collects samples of soil, air, groundwater, and/or building materials to analyze for quantitative values of various contaminants. Conclusions are drawn from data collected on site, as well as from regional and local information available. If conditions are identified that may require remediation, the recommendations section of the report will discuss general options available.

Related Credits

Teams choosing to develop a brownfield site may also be eligible for SSc1 Site Selection.

SSc4 ALTERNATIVE TRANSPORTATION—INTRODUCTION

Reducing dependence on automobiles can help lower the energy demand of transportation and the associated greenhouse gases, as well as decrease the need for, or size of, parking lots and, in return, allow more area for outdoor open space. It will also reduce the pressure on cities and states to build or increase the size of roads and expressways. The reduction in impervious surfaces also reduces runoff, lowers demands on existing storm sewers, and contributes to reducing heat island effect.

"Transportation sources accounted for 29 percent of total U.S. greenhouse gas (GHG) emissions in 2006. Transportation is the fastest-growing source of GHGs in the U.S., accounting for 47 percent of the net increase in total U.S. emissions since 1990. Transportation is also the largest end-use source of CO_2, which is the most prevalent greenhouse gas. These estimates of transportation GHGs do not include emissions from additional life-cycle processes, such as the extraction and refining of fuel and the manufacture of vehicles, which are also a significant source of domestic and international GHG emissions."[3]

The following summarized series of credits seek to decrease automobile use to reduce pollution and land development impacts. The credits encourage alternate modes of transportation that, if implemented, not only help to reduce GHG, but also provide health, community, and social benefits.

- SSc4.1 Alternative Transportation—Public Transportation Access
- SSc4.2 Alternative Transportation—Bicycle Storage and Changing Rooms
- SSc4.3 Alternative Transportation—Low-Emitting and Fuel-Efficient Vehicles
- SSc4.4 Alternative Transportation—Parking Capacity

Exemplary Performance

One Innovation in Design point for exemplary performance is available to teams that, by using any combination of SSc4 Alternative Transportation credits, establish a comprehensive management plan that demonstrates a quantifiable reduction in personal automobile use.

SSc4.1 ALTERNATIVE TRANSPORTATION—PUBLIC TRANSPORTATION ACCESS

Intent

Proximity to mass transit is viewed as a benefit to many building occupants and can help businesses by attracting and retaining employees and customers. It can reduce commuting costs and provide transportation to those without other modes of transportation.

When deciding if this credit is appropriate for the project, teams should determine if mass transportation meets the needs of the building's occupants. Existing transportation networks should be used before creating new lines of transportation. Safe, direct sidewalks and paths from the project to the mode of transportation should also be provided.

Requirements

To attain this credit, the project must be located within ½ mile of a railway station, or within ¼ mile of two bus line stops, as a pedestrian would walk. Railways include commuter rail, light rail, or subway. To qualify for this credit, the rail stop must be existing or planned and funded. To demonstrate compliance with this credit, teams should develop a drawing, aerial photograph, or map to illustrate the walking distance to transit stops.

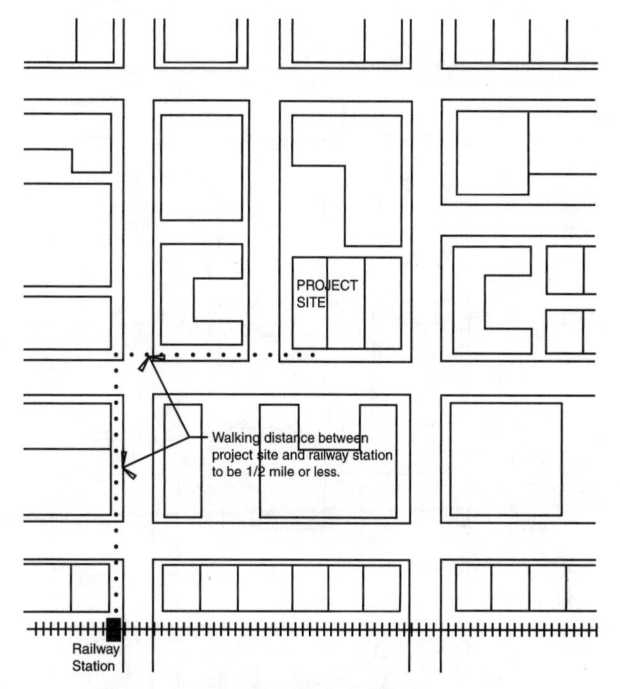

Figure 2.3 Public Transportation Access—Railway

If private shuttle buses are used to satisfy the credit requirements, they must operate during the most frequent commuting hours for employees of the facility, and connect to public transportation.

Selecting a site with proximity to existing transit infrastructure, and developing a site vicinity plan reflecting the locations and types of nearby mass transit, can help teams determine if mass transit is suitable for their project. The site vicinity plan can be part of a larger transportation management plan for a comprehensive approach to address all of the Alternative Transportation credits.

Exemplary Performance

One Innovation in Design point for exemplary performance is available to teams that double transit ridership by locating the project (1) within ½ mile of at least two existing commuter rail, light rail, or subway lines; or (2) within ¼ mile of at least two stops for four or more bus lines.

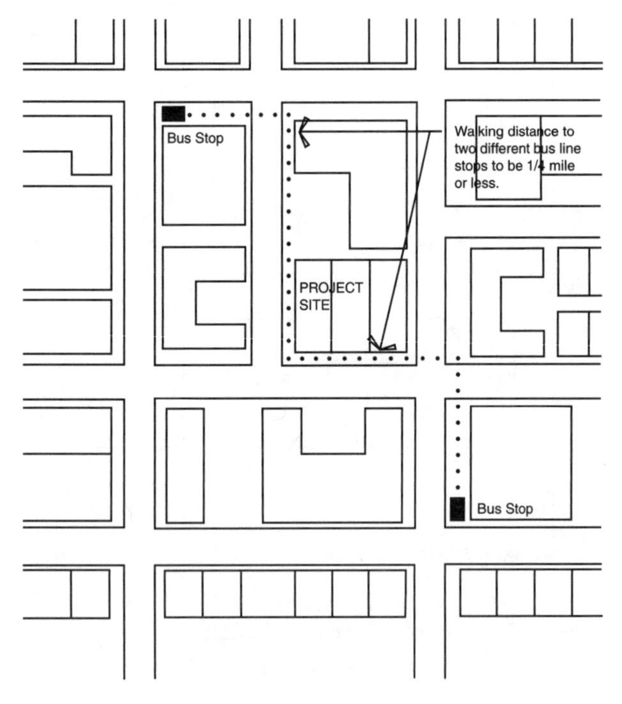

Figure 2.4 Public Transportation Access—Bus

Related Credits

Project sites with access to public transportation are likely to be situated in an urban location and may contribute to the requirements of SSc1 Site Selection and SSc2 Development Density and Community Connectivity.

SSc4.2 ALTERNATIVE TRANSPORTATION—BICYCLE STORAGE AND CHANGING ROOMS

Intent

"In communities across the world, there is a growing need and responsibility to provide options that give people the opportunity to bike—to bike more often, to bike to more places, and to feel safe while doing so. The benefits of riding a bicycle—whether for utilitarian or recreational purposes—can be expressed in terms of improved environmental and personal health, reduced traffic congestion, enhanced quality of life, economic rewards, as well as others."[4]

"Bicycle commuting produces no emissions, has zero demand for petroleum-based fuels, relieves traffic congestion, reduces noise pollution, and requires far less infrastructure for roadways and parking lots. Roadways and parking lots, on the other hand, produce stormwater runoff, contribute to the urban heat island effect, and encroach on green space."[5]

Providing bicycle storage can be as simple as installing exterior bicycle racks, which are relatively inexpensive and require only a fraction of the space to park the same number of cars. Convenient and safe access to year-round bicycle racks and showers should be provided from bicycle paths that are functional, and that offer direct connections to public bicycle paths and streets appropriate for bicycle riding.

Requirements

Teams pursuing this credit for commercial and institutional projects must provide bicycle storage for 5 percent or more of the building users, as well as shower and changing facilities in the building or within 200 yards of the building entrance. Residential projects must provide bicycle storage for 15 percent of building occupants. Shower facilities are not required, but bicycle storage must be sheltered from the elements.

Related Credits

The materials used to create on-site bicycle lanes can impact site drainage and heat island effect. If low-albedo, permeable surfaces are used, they can help to achieve SSc6 Stormwater Design and SSc7.1 Heat Island Effect—Nonroof.

SSc4.3 ALTERNATIVE TRANSPORTATION—LOW-EMITTING AND FUEL-EFFICIENT VEHICLES

Intent

Using mass transit, walking, or bicycling is not an option for commuting to and from work for everyone. Carpooling or using an alternative fuel or fuel-efficient vehicle can be sensible alternatives. These options offer the possibility of producing fewer air pollutants from vehicle travel and reduce the negative effects of gasoline production.

Electric fuel vehicles do not produce greenhouse gases (GHG) during operation, but GHGs are produced during the manufacturing of the electrical power they use. The amount of GHGs produced by the creation of electricity varies depending on the method used to produce the electricity. For example, a hydroelectric plant will produce less GHG than a coal-fired electrical plant.

Requirements

Project teams have four options to attain this credit:

Option 1: Supply preferred parking for low-emitting and fuel-efficient vehicles for 5 percent of the total parking capacity of the site, or provide parking for all low-emitting and fuel-efficient vehicles discounted at least 20 percent.

Option 2: Supply alternative fuel refueling stations for 3 percent of the total parking capacity of the site.

Option 3: Supply low-emitting and fuel-efficient vehicles and preferred parking for 3 percent of Full Time Equivalent (FTE) building occupants.

Option 4: Provide building occupants access to low-emitting and fuel-efficient vehicle-sharing programs.

Related Credits

Projects that provide preferred parking without increasing parking capacity may be able to earn SSc4.4 Alternative Transportation—Parking Capacity.

SSc4.4 ALTERNATIVE TRANSPORTATION—PARKING CAPACITY

Intent

This credit encourages the utilization of alternative forms of transportation by limiting on-site parking. This approach can be effective if sufficient forms of alternative transportation that fit the needs of the building occupants are available. If the available forms of mass transit do not fit the needs of the building occupants and the available on-site parking is limited, it can result in overflow parking to city streets, potentially causing the neighborhood residents to respond negatively. Sufficient study and consideration should be given to the required parking capacity early in the planning phases of the project to make sure the prevailing zoning codes are met and, if possible, minimize the parking spaces provided.

Impervious parking lot surfaces create negative environmental impacts, such as increasing heat island effect and stormwater runoff. Promoting carpooling and restricting parking lot size can increase open space and reduce runoff and heat island effect.

Requirements

Non-residential projects can attain this credit by achieving one of the following three requirement options:

Option 1: Parking capacity does not exceed minimum zoning requirements, and 5 percent of parking is preferred parking for High Occupancy Vehicles (HOVs).

Option 2: If parking is less than 5 percent of FTE building occupants, then 5 percent of the total parking spaces are to be identified for HOVs must be preferred parking.

Option 3: Do not provide new parking.

Residential and mixed-use residential with commercial/retail projects can attain this credit by designing parking capacity not to exceed minimum local zoning requirements, and by providing infrastructure and support programs facilitating mass transit shuttle services, HOV parking, car-share services and carpool drop-off areas, or by not providing new parking.

Referenced Standard

The referenced standard for this credit is the Institute of Transportation Engineers, Parking Generation Study, 2003. This standard can be downloaded at *www.ite.org*.

Related Credits

Minimizing parking by not exceeding zoning requirements can increase the opportunity for open space, minimize heat island effect and stormwater runoff, which can contribute to SSc5.1 Site Development—Protect or Restore Habitat, SSc5.2 Site Development—Maximize Open Space, SSc6 Stormwater Design and/or SSc7.1 Heat Island Effect—Nonroof.

SSc5.1 SITE DEVELOPMENT—PROTECT OR RESTORE HABITAT

Intent

Preservation and protection of natural areas during construction is the most ecologically responsible approach to maintain the natural resources of the site during construction.

If this is not possible, the next best option is to restore damaged areas. During construction, the establishment and implementation of structural site boundaries and staging areas can be very effective methods to reduce damage to the site and help to preserve wildlife habitats and migration corridors. Efforts made during construction to maintain existing natural areas can also prove to be cost-effective landscaping strategies. Trees, plants, and other landscaping grown off-site may not survive transplanting. Creating new site amenities, such as water features, can add significant cost to a project. Using native or adapted planting can reduce cost over the life-cycle of the building because they require less fertilizer, pesticides, irrigation, and maintenance than non-native plants. These issues can be avoided by maintaining existing natural site features and landscaping.

Natural areas support indigenous plants and regional animal populations, which allow for natural management of stormwater volumes, water bodies, exposed rock, bare ground, and other features.

Construction documents should note site protection requirements and clearly identify construction site disturbance boundaries. Before site work begins, contractors should identify lay-down, recycling, and disposal areas. Existing trees that are to remain should be protected with construction fencing at the dripline to protect them from damage. Paved areas should be used for staging activities. Installation of infrastructure should also be coordinated to minimize site disruption.

Requirements

Greenfield sites can achieve this credit by limiting site disturbance, including earthwork and clearing of vegetation, to (1) 40 feet beyond the building perimeter; (2) 10 feet beyond surface walkways, parking patios, and utilities that are smaller than 12 inches in diameter; (3) 15 feet beyond primary roadway curbs, walkways, and main utility branch trenches; and (4) 25 feet beyond constructed areas with permeable surfaces that require additional staging areas in order to limit compaction in the paved area.

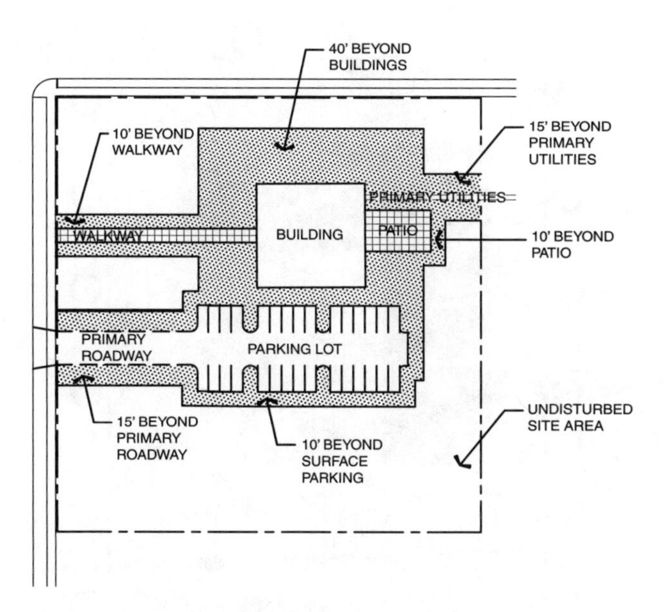

Figure 2.5 Site Disturbance Boundaries Sample Plan

Previously developed or graded sites can achieve this credit by restoring a minimum of 50 percent of the site, excluding the building footprint, by planting native or adapted vegetation. If the project is earning SSc2 Development Density and Community Connectivity and using a vegetated roof, the area of the vegetated roof can be applied to this calculation, if plants are native and/or adapted.

Exemplary Performance

One Innovation in Design point for exemplary performance is available to teams that restore or protect

- a minimum of 75 percent of the site, excluding the building footprint; or
- 30 percent of the total site, including the building footprint, with native or adapted vegetation, whichever is greater.

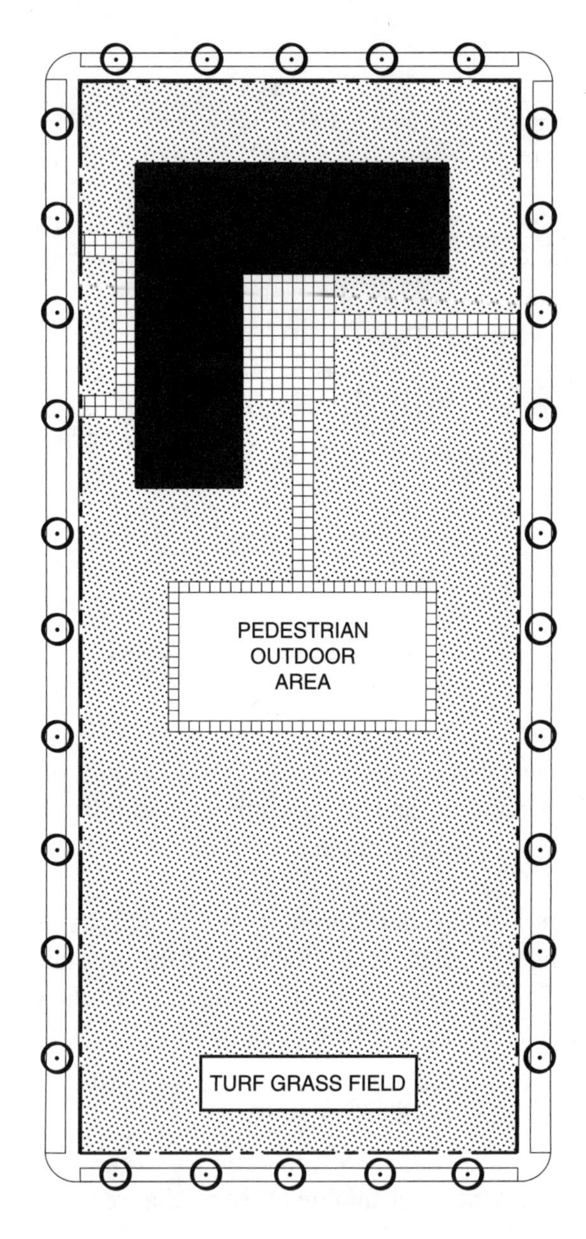

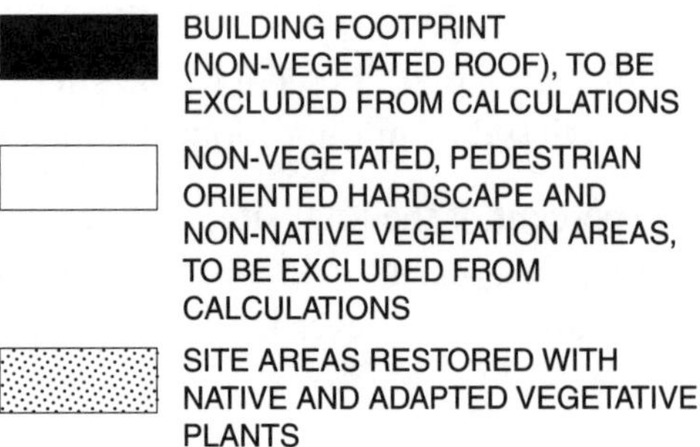

Figure 2.6 Natural Area Sample Plan

Related Credits

Teams choosing to protect or restore habitat can increase open space on the site, which minimizes the stormwater runoff and heat island problems that stem from impervious surfaces. Use of native vegetation on site or as part of a vegetated roof may contribute to achieving the following credits:

- Sc5.2 Site Development—Maximize Open Space
- SSc6.1 Stormwater Design—Quantity Control
- SSc6.2 Stormwater Design—Quality Control
- SSc7.1 Heat Island Effect—Nonroof
- SSc7.2 Heat Island Effect—Roof
- WEc1 Water Efficient Landscaping

SSc5.2 SITE DEVELOPMENT—MAXIMIZE OPEN SPACE

Intent

Many city zoning requirements stipulate a minimum square footage of open space to be provided. This requirement not only ensures that buildings are not constructed from lot line to lot line, it also creates space for vegetation and wildlife. Open space can also reduce heat island effect, increase stormwater infiltration, and provide humans the opportunity to connect to the outdoors, as well as potentially increase property values. Even small spaces can provide habitats for wildlife populations and add property value. The intent of this credit is to develop the building footprint to provide a high percentage of open space to encourage biodiversity.

Minimizing the building footprint does not necessarily mean minimizing the building square footage. Increasing the building height can reduce the building footprint and provide the opportunity to increase site open space.

When choosing the location and shape of the development footprint, teams should consider sustainable site issues, including the existing ecosystem of the site, building orientation, daylighting, heat island effect, stormwater generation, existing vegetation, and existing green corridors. Connections to green corridors and open land can be preserved by designing compact parking lots and a building footprint oriented to minimize impact to the site.

If a green roof is used to fulfill the open space requirement, an added benefit can be realized through reduced energy costs due to the insulating properties of the green roof.

Requirements

Sites with local zoning open space requirements can achieve this credit by reducing the development footprint and/or providing vegetated open space within the project boundary to exceed the local zoning's open space requirement for the site by 25 percent.

Sites with no local zoning requirements can achieve this credit by designating open space area adjacent to the building that is equal to the development footprint.

Sites that have a zoning ordinance that does not provide for open space can achieve this credit by establishing an open space equaling 20 percent of the project site area.

In all cases noted above, if the site is in an urban area and is earning SSc2 Development Density and Community Connectivity, the vegetated roof areas and pedestrian oriented hardscape with a minimum 25 percent vegetated open space can contribute to attaining the credit. Wetlands or naturally designed ponds may count as open space if the vertical:horizontal side slope gradient average 1:4 is vegetated.

Exemplary Performance

One Innovation in Design point for exemplary performance is available to teams that demonstrate they have doubled the amount of required open space. All designated open space must be within the LEED® project boundary.

Related Credits

Providing vegetated open space on the project site may contribute to stormwater mitigation goals, reduce the urban heat island effect, and help project teams earn the following credits:

- SSc6.1 Stormwater Design—Quantity Control
- SSc6.2 Stormwater Design—Quality Control
- SSc7.1 Heat Island Effect—Nonroof
- SSc7.2 Heat Island Effect—Roof

SSc6.1 STORMWATER DESIGN—QUANTITY CONTROL

Intent

"Stormwater has been identified as one of the leading sources of pollution for all waterbody types in the United States. Further, the impacts of stormwater pollution are not static: they usually increase with more development and urbanization."[2] The intent of this credit is to reduce impervious ground cover to increase on-site infiltrations, manage stormwater runoff, and eliminate contaminants to limit disruption of natural hydrology. Reducing the water flowing into municipal systems lowers the quantity of water requiring treatment and decreases infrastructure expansion and maintenance.

Teams can maintain natural stormwater flows and promote infiltration by specifying vegetated roofs, pervious paving, and other methods to minimize impervious surfaces to prevent or delay stormwater runoff. Stormwater volumes captured on-site can also be used for nonpotable purposes, such as landscape irrigation, toilet and urinal flushing, and custodial use.

Identifying natural drainage concepts at the beginning of site planning allows the design team sufficient time to economically integrate them into the site amenities. Compaction of the site by adding impervious surfaces such as roads, walkways, and parking lots increases runoff, which can contain contaminants from the surfaces the stormwater flows across. Contaminants can include pesticides, fertilizers, vehicle fluids, mechanical equipment wastes, and a variety of other chemicals that cause damage to receiving bodies of water.

Strategies that teams use to achieve this credit will vary depending on conditions of the site. If the site is undeveloped, teams will want to preserve the existing stormwater flows, the natural soil conditions, and the habitat and rainfall characteristics. If a site has been previously developed, the project team will design the site to improve stormwater management and restore the natural functions of the site.

Requirements

To achieve this credit for sites with existing imperviousness of 50 percent or less, teams must

- implement a stormwater management plan to ensure that post-development peak discharge rate and quantity does not exceed pre-development peak discharge rate and quantity for 1- and 2-year, 24-hour design storms; or
- execute receiving stream channel protection from excessive erosion. This plan must include a description of the quantity control strategies to protect receiving stream channels.

For sites with existing imperviousness of greater than 50 percent, teams must execute a stormwater management plan to reduce the volume of stormwater runoff by 25 percent as compared to a 2-year, 24-hour design storm.

Exemplary Performance

A standardized exemplary performance option has not been established for this credit, but teams can apply for exemplary performance by documenting a comprehensive approach to capture and treat stormwater runoff. Teams must demonstrate performance greatly exceeding the credit threshold. Only one exemplary performance point can be achieved between SSc6.1 Stormwater Design—Quantity Control and SSc6.2 Stormwater Design—Quality Control.

Related Credits

Restoring the rate and quantity of stormwater runoff will result in increased on-site infiltration, reduce the amount of stormwater treatment needed, and can help meet the requirements of SSc6.2 Stormwater Design—Quality Control.

Reduction in impervious surfaces by using pervious pavement, vegetated roofing, and vegetated open space can help meet the requirements of the following credits:

- SSc5.1 Site Development—Protect or Restore Habitat
- SSc5.2 Site Development—Maximize Open Space

- SSc7.1 Heat Island Effect—Nonroof
- SSc7.2 Heat Island Effect—Roof

Reducing stormwater runoff by harvesting it for nonpotable uses can help teams pursuing WEc1 Water Efficient Landscaping and WEc3 Water Use Reduction.

SSc6.2 STORMWATER DESIGN—QUANTITY CONTROL

Intent

Conveyance and treatment of stormwater requires significant municipal infrastructure and maintenance. The intent of this credit is to manage stormwater runoff to reduce or eliminate water pollution through the reduction of impervious surfaces, increased on-site infiltration, and the removal of pollutants from stormwater runoff. The best way to achieve this goal is to keep all stormwater on site. If the site doesn't have excellent percolation or other means to prevent stormwater from leaving the site, reducing the quantity and slowing the rate of discharge can be achieved by decreasing the amount of impervious area to maintain natural stormwater flows and promote infiltration. Strategies to achieve this credit include the following:

- Providing vegetative swales for stormwater runoff instead of structured pipes
- Creating water ponds or retention areas to temporarily store or delay stormwater runoff
- Designing mechanical and natural treatment for site stormwater

Teams can use any combination of structural and nonstructural stormwater management measures to minimize or mitigate impervious area to achieve this credit. Nonstructural stormwater management methods include rain gardens, vegetated swales, and infiltration to capture and treat runoff by allowing stormwater to naturally filter into the soil. Nonstructural methods of stormwater management are often preferred because they are less expensive to implement and help to recharge natural aquifers. Structural stormwater management techniques include cisterns, manhole treatment devices, and ponds. Structural control can be preferred on sites with limited space because it can effectively clean stormwater.

The stormwater management plan should include a description of the Best Management Practices (BMPs) used on the project. The plan should describe the methods used to capture and treat stormwater runoff and how they contribute to reducing imperviousness and/or increasing infiltration. The plan should also document how 90 percent of the average annual rainfall is captured and treated. For this credit, 90 percent of the average rainfall is equivalent to treating runoff from the amounts established in the following table.

Watershed	Rainfall per 24 Hours (ins)
Humid	1
Semiarid	0.75
Arid	0.5

From LEED® 2009 Reference Guide

Requirements

Teams can achieve this credit by using acceptable BMPs to implement a stormwater management plan to reduce impervious ground cover, promote infiltration, and capture and treat stormwater runoff equaling 90 percent average annual rainfall. The stormwater treatment system must remove 80 percent Total Suspended Solids (TSS) and 40 percent Total Phosphorus (TP), based on a 2-year, 24-hour storm.

This credit can also be achieved if in-field performance monitoring data exist that demonstrates compliance with the credit criteria. The data must be compliant with the accepted protocol Technology Acceptance Reciprocity Partnership (TARP).

Related Credits

Teams using BMPs to capture and treat stormwater runoff also contribute to the reduction of the runoff volume, which can assist in attaining SSc6.1 Stormwater Design—Quantity Control.

Teams choosing to reduce impervious surfaces by providing pervious pavements, vegetated roofing, and/or vegetated open space can contribute to attaining the following credits:

- SSc5.1 Site Development—Protect or Restore Habitat
- SSc5.2 Site Development—Maximize Open Space
- SSc7.1 Heat Island Effect—Nonroof
- SSc7.2 Heat Island Effect—Roof

Teams using rain gardens, vegetated swales, or rainwater harvesting systems to reduce or eliminate the need for landscape irrigation my also be able to achieve WEc1 Water Efficient Landscaping.

SSc7.1 HEAT ISLAND EFFECT—NONROOF

Intent

"The term 'heat island' describes built up areas that are hotter than nearby rural areas. The annual mean air temperature of a city with 1 million people or more can be 1.8–5.4°F (1–3°C) warmer than its surroundings. In the evening, the difference can be as high as 22°F (12°C). Heat islands can affect communities by increasing summertime peak energy demand, air conditioning costs, air pollution and greenhouse gas emissions, heat-related illness and mortality, and water quality."[6] Minimizing the impact of heat islands on microclimates and human and wildlife habitats by reducing the gradient between developed and undeveloped areas is the intent of this credit. Shading roofs, roads, sidewalks, and other constructed surfaces with landscape features, replacing hard surfaces with permeable surfaces such as green roofs, installing open-grid paving, and using high-albedo materials are good strategies to mitigate heat island effect. Combining the strategies can increase their effectiveness.

The Solar Reflectance Index (SRI) of hard paving surfaces can vary depending on their characteristics. A dark material will reflect less light than a light-colored material. The higher the SRI number, the more light is reflected.

Requirements

Teams can achieve this credit using any combination of the following options for 50 percent of the site hardscape:

- Provide shade from existing tree canopy within five years of landscape irrigation, with the source of shade installed at time of occupancy.
- Provide shade from solar panels that produce energy used to offset nonrenewable resource use.
- Use paving materials with a minimum SRI of 29.
- Use an open-grid paving system that is at least 50 percent impervious.

Teams can also achieve this credit if their design requires at least 50 percent of parking spaces to be covered. Roofs used to shade parking must have a minimum SRI of 29.

Exemplary Performance

One Innovation in Design point for exemplary performance is available to teams that demonstrate that either (1) 100 percent of nonroof impervious surfaces have been constructed with high-albedo materials or open-grid paving, or will be shaded within five years; or (2)100 percent of the on-site parking spaces must be located under cover.

Referenced Standards

There are five referenced standards for this credit, each addressing a different aspect of this credit's requirements.

1. ASTM Standard E408-71(1996)e1 —Standard Test Methods for Total Normal Emittance of Surfaces Using Inspection-Meter Techniques. This standard states: "These test methods cover determination of the total normal emittance (Note) of surfaces by means of portable, inspection-meter instruments. Note, total normal emittance (N) is defined as the ratio of the normal radiance of a specimen to that of a blackbody radiator at the same temperature. The equation relating N to wavelength and spectral normal emittance. These test methods are intended for measurements on large surfaces when rapid measurements must be made and where a nondestructive test is desired. They are particularly useful for production control tests."[1]
2. ASTM C1371-04a. This standard states: "This test method covers a technique for determination of the emittance of typical materials using a portable differential thermopile emissometer. The purpose of the test method is to provide a comparative means of quantifying the emittance of opaque, highly thermally conductive materials near room temperature as a parameter in evaluating temperatures, heat flows, and derived thermal resistances of materials."[1]
3. ASTM E903-96—Standard Test Method for Solar Absorptance, Reflectance, and Transmittance of Materials Using Integrating Spheres. This standard states: "This test method covers the measurement of spectral absorptance, reflectance, and transmittance of materials using spectrophotometers equipped with integrating spheres."[1]
4. ASTM Standard E1918-97—Standard Test Method or Measuring Solar Reflectance of Horizontal and Low-Sloped Surfaced in the Field. "This test method covers the measurement of solar reflectance of various horizontal and low-sloped surfaces and materials in the field, using a pyranometer. The test method is intended for use when the sun angle to the normal from a surface is less than 45."[1]
5. ASTM Standard C1549-04—Standard Test Method for Determination of Solar Reflectance Near Ambient Temperature Using a Portable Solar Reflectometer. This standard states: "This test method covers a technique for determining the solar reflectance of flat opaque materials in a laboratory or in the field using a commercial portable solar reflectometer. The purpose of the test method is to provide solar reflectance data required to evaluate temperatures and heat flows across surfaces exposed to solar radiation."[1]

Related Credits

Teams choosing to locate parking underground can limit site disturbance and maximize open space, which can help the team achieve SSc5.2 Site Development—Maximize Open Space.

Teams using open-grid pavement can capture and treat stormwater runoff, which can assist in achieving SSc6.1 Stormwater Design—Quantity Control and SSc6.2 Stormwater Design—Quality Control.

Potable water for landscape irrigation can be reduced if the project team chooses to use vegetation to shade hardscapes, which can help the team achieve WEc1 Water Efficient Landscaping.

SSc7.2 HEAT ISLAND EFFECT—ROOF

Intent

Teams can mitigate heat island effect associated with roofs by installing Energy Star® compliant roofing products and/or vegetative roof systems. Using either of these strategies or a combination of the two can effectively address the intent of this credit to minimize the impact of heat islands on microclimates, human, and wildlife habitats by reducing the gradient between developed and undeveloped areas.

Vegetative roofs do have a higher initial cost for structural support and materials; however, in addition to decreasing stormwater runoff, they also provide reduced energy cost by replacing heat-absorbing surfaces with plants that cool the air through evapotranspiration. Vegetative roofs also benefit the building by providing additional insulation and protection of the roof membrane.

Roofs comprised of high reflectance materials can minimize heat island effect and save energy. The higher the SRI value, the more light is reflected.

When designing a vegetative roof system, teams can choose from may options. They can be installed as a complete or modular system. Some vegetative roofs are designed to be enjoyed as planting beds or gardens, others are designed to be utilitarian only and have little or no human interaction. Both types of systems offer the same basic qualities: they provide insulation benefits, retain stormwater, have longer lives, and require less maintenance then conventional roof systems.

Requirements

Teams using reflective roof materials can achieve this credit by designing 75 percent of roof surfaces to be high-albedo. Low-sloped roofs must have an SRI of at least 78 and steep-sloped roofs must have an SRI of at least 29.

Projects using a green roof can achieve this credit if 50 percent of the roof is vegetated.

If a team chooses to combine reflective and green roof systems, they can achieve this credit if the combined roof areas of high-albedo and vegetation meet the following formula:

(Area of Roof meeting minimum SRI / 0.75) + (Area of Vegetated Roof / 0.50) ≥ Total Roof Area

Exemplary Performance

One Innovation in Design point for exemplary performance is available to teams that demonstrate that 100 percent of the project's roof area is a vegetative roof system. Areas where mechanical equipment, photovoltaic panels and skylights are not included.

Referenced Standards

There are six referenced standards for this credit, each addressing a different aspect of this credit's requirements.

1. ASTM Standard E1980-01—Practice for Calculating Solar Reflectance Index

of Horizontal and Low-Sloped Opaque Surfaces. This standard states: "This practice covers the calculation of the Solar Reflectance Index (SRI) of horizontal and low-sloped opaque surfaces at standard conditions. The method is intended to calculate SRI for surfaces with emissivity greater than 0.1."[1]

2. ASTM Standard E408-71(1996)e1—Standard Test Methods for Total Normal Emittance of Surfaces Using Inspection-Meter Techniques. This standard states: "These test methods cover determination of the total normal emittance (Note) of surfaces by means of portable, inspection-meter instruments. Note, total normal emittance (N) is defined as the ratio of the normal radiance of a specimen to that of a blackbody radiator at the same temperature. The equation relating N to wavelength and spectral normal emittance. These test methods are intended for measurements on large surfaces when rapid measurements must be made and where a nondestructive test is desired. They are particularly useful for production control tests."[1]
3. ASTM E903-96—Standard Test Method for Solar Absorptance, Reflectance, and Transmittance of Materials Using Integrating Spheres. This standard states: "This test method covers the measurement of spectral absorptance, reflectance, and transmittance of materials using spectrophotometers equipped with integrating spheres."[1]
4. ASTM Standard E1918-97—Standard Test Method or Measuring Solar Reflectance of Horizontal and Low-Sloped Surfaced in the Field. This standard states: "This test method covers the measurement of solar reflectance of various horizontal and low-sloped surfaces and materials in the field, using a pyranometer. The test method is intended for use when the sun angle to the normal from a surface is less than 45."[1]
5. ASTM Standard C1371-04—Standard Test Method for Determination of Emittance of Material Near Room Temperature using Portable Emissometers. This standard states: "This test method covers a technique for determination of the emittance of typical materials using a portable differential thermopile emissometer. The purpose of the test method is to provide a comparative means of quantifying the emittance of opaque, highly thermally conductive materials near room temperature as a parameter in evaluating temperatures, heat flows, and derived thermal resistances of materials."[1]
6. ASTM Standard C1549-04—Standard Test Method for Determination of Solar Reflectance Near Ambient Temperature Using a Portable Solar Reflectometer. This standard states: "This test method covers a technique for determining the solar reflectance of flat opaque materials in a laboratory or in the field using a commercial portable solar reflectometer. The purpose of the test method is to provide solar reflectance data required to evaluate temperatures and heat flows across surfaces exposed to solar radiation."[1]

Related Credits

Project teams that choose to install a green roof are able to provide open space and habitat, as well as capture and treat stormwater, which can help them achieve the following credits:

- SSc5.1 Site Development—Protect or Restore Habitat
- SSc5.2 Site Development—Maximize Open Space

- SSc6.1 Stormwater Design—Quantity Control
- SSc6.2 Stormwater Design—Quality Control

If a highly reflective roof combined with a vegetative roof is designed for the project, this may assist teams pursuing EAc1 Optimized Energy Performance by helping to decrease cooling loads.

SSc8 LIGHT POLLUTION REDUCTION

Intent

Light pollution can obscure the stars in the night sky, interfere with astronomical observatories, disrupt ecosystems and cause adverse health effects. There are two main types of light pollution: the first is light that intrudes on natural or low-light settings, and the second is excessive light that leads to discomfort and other health problems. The intent of this credit is to minimize building and site light trespass, reduce glare in order to increase nighttime visibility and reduce the impact on nocturnal environments.

Lighting only areas necessary for safety and comfort, providing only the light levels necessary to meet the design intent, and selecting efficient fixtures, will minimize light pollution and save energy.

Requirements

To achieve this credit, teams must fulfill one of two options for interior lighting, and one option for exterior lighting.

For interior lighting, teams can either control lighting between 11:00 PM and 5:00 AM by using automatic lighting controls to reduce lighting power by 50 percent for all non-emergency light fixtures with a direct line of site to any building opening. For after-hours lighting, manual controls or occupant sensors must be installed to allow lighting to be turned on for intervals of no more than 30 minutes. Or, for the same period of time, all openings in the building envelope with a direct line of sight to any non-emergency luminaire, must have a transmittance of less than 10 percent, which can be achieved by using automatic controls for shielding the opening.

For exterior lighting, the team must only light areas as required for safety and comfort. Follow ANSI/ASHRAE/IESNA Standard 90.1—2007, with errata, but without addendum. All projects must be classified as a specific zone according to IESNA RP-33: LZ1, LZ2, LZ3 or LZ4 as defined in the following list.

IESNA RP-33 Lighting Zones:

- LZ1—Dark, Park and Rural Areas
- LZ2—Low, Residential Areas
- LZ3—Medium, Commercial/Industrial, High-Density Residential Areas
- LZ4—High, Major City Centers, Entertainment Districts

Referenced Standard

The referenced standard for this credit is ANSI/ASHRAE/IESNA 90.1—2007, lighting section 9, without amendments (Energy Standard for Buildings except low-rise residential).

Related Credits

Energy savings beyond meeting ANSI/ASHRAE/IESNA Standard 90.1—2007 can be applied to attaining EAc1 Optimized Energy Performance.

ABBREVIATIONS AND ACRONYMS

ACEEE	American Council for an Energy-Efficient Economy
AFV	Alternative Fuel Vehicles
ASHRAE	American Society of Heating and Refrigeration Engineers of North America
ASTM	American Society of Testing and Materials
BMP	Best Management Practices
CGP	Construction General Permit
ESA	Environmental Site Assessment
ESC	Erosion and Settlement Control Plan
GHG	Green House Gases
HOV	High Occupancy Vehicle
IESNA	Illuminating Engineering Society of North America
LPD	Lighting Power Density
NFIP	National Flood Insurance Program
NPDES	National Pollutant Discharge Elimination Program
NRDC	National Resources Defense Council
SF	Square Footage
SRI	Solar Reflectance Index
SWPP	Stormwater Pollution Prevention Plan
TARP	Technology Acceptance Reciprocity Partnership
TIF	Tax Increment Financing
TP	Total Phosphorous
TSS	Total Suspended Solids
UST	Underground Storage Tank
ZEV	Zero Emission Vehicles

FOOTNOTES

1. U.S. Environmental Protection Agency, *Erosion & Sediment Control*, 2009, http://www.epa.gov/nps/ordinance/erosion.htm.
2. U.S. Environmental Protection Agency, *Reducing Stormwater Costs through Low Impact Development (LID) Strategies and Practices,* 2007, 1. http://www.epa.gov/owow/nps/lid/costs07/documents/reducingstormwatercosts.pdf.
3. U.S. Environmental Protection Agency, *Transportation and Climate*, 2009, http://www.epa.gov/otaq/climate/index.htm.
4. Pedestrian and Bicycle Information Center, *Why Bicycle?*, 2009, http://www.bicyclinginfo.org/why/.
5. *LEED® Reference Guide for Green Building Design and Construction*, 2009, 51. http://www.usgbc.org/Store/PublicationsList_New.aspx?CMSPageID=1518&gclid=CO3_oeGR4pwCFWBB5god1WcBKQ
6. U.S. Environmental Protection Agency, *Heat Island Effect*, 2009, http://www.epa.gov/heatisland/index.htm.

LESSON 2 QUIZ

1. Of the following, which three are the major goals of SSp1 Construction Activity Pollution Prevention?

   A. Prevent air pollution from dust and particulate matter during construction.

   B. Prevent air pollution from CO_2 released during construction.

   C. Limit sedimentation in storm sewers and nearby streams during construction.

   D. Avoid soil loss due to wind and stormwater erosion during construction.

   E. Identify and remove Underground Storage Tanks (USTs) at the construction site.

2. A publishing company has chosen to construct a 18,000 square foot building on a site designated as a brownfield by the U.S. EPA. The site is located in a well-developed urban area with access to public transportation and amenities. Which three of the following credits should the team pursue?

   A. SSc1 Site Selection

   B. SSc2 Development Density and Community Connectivity

   C. SSc3 Brownfield Redevelopment

   D. SSc4 Alternative Transportation

   E. SSc5 Site Development

3. What are the methods described in the 2003 EPA Construction General Permit to control erosion and sedimentation? (Choose two.)

   A. Minimization

   B. Structural control

   C. Sewerage drainage

   D. Stabilization

   E. Balance

4. Sustainable Sites Prerequisite 1 requires the implementation of an erosion and sedimentation plan. Which three of the following objectives must be described in the plan?

   A. Prevention of sedimentation of storm sewers of receiving streams

   B. Prevention of loss of turf grass during construction

   C. Prevention of loss of soil during construction by stormwater runoff and/or wind erosion, including protecting topsoil by stockpiling for reuse

   D. Prevention of pollution of the air with dust and particulate matter

   E. Protection of existing trees during construction

5. Which three of the following are required to be achieved under SSc1 Site Selection?

   A. Do not develop on land within 100 feet of a wetland.

   B. Do not develop on prime farmland.

   C. Do not develop on land defined as a virgin forest.

   D. Do not develop on land lower than five feet below the 100-year flood plane.

   E. Do develop on a brownfield site.

   F. Do not develop on land lower than five feet above the 100-year flood plane.

6. Which three of the following are site disturbance parameters for SSc5.1 Site Development—Protect or Restore Habitat?

   A. 15 feet beyond all water features

   B. 10 feet beyond surface walkways, patios, surface parking, and utilities less than 12 inches in diameter

   C. 10 feet beyond all trees

   D. 15 feet beyond primary roadway curbs and main utility branch trenches

   E. 40 feet beyond building parameters

7. Which agency provides confirmation that a site is classified as a brownfield?

   A. Local building department

   B. USGBC

   C. Local, state, or federal agency

   D. FDA

   E. ASHRAE

8. Decreasing permeability increases which three of the following stormwater run-off properties?

   A. Velocity

   B. Measurable run-off

   C. Percolation

   D. Volume

   E. Porosity

9. What percentage of bicycle parking is required to be provided in SSc4.2 Alternative Transportation—Bicycle Storage and Changing Rooms?

   A. 5 percent FTE for commercial buildings and 15 percent for residential buildings

   B. 15 percent FTE for commercial buildings and 5 percent for residential buildings

   C. 15 percent FTE for commercial buildings and 15 percent for residential buildings

10. What does the term "in situ remediation" refer to?

   A. Solid contamination that must be removed from the site for treatment

   B. Liquid contamination that must be removed from the site for treatment

   C. Solid or liquid contamination that must be removed from the site for treatment

   D. Solid contamination that is treated on site

   E. Solid or liquid contamination that is treated on site

QUIZ ANSWERS

Lesson 2

1. **A, C, D** CO_2 reduced during construction and USTs are not addressed in SSp1.
2. **A, C, D** The site is a brownfield, located in a well-developed urban area near public transportation.
3. **B, D** Only structural control and stabilization are described in the 2003 EPA Construction General Permit to control erosion and sedimentation.
4. **A, B, C** See the requirements for the credit for prevention of sedimentation of storm sewers of receiving streams. Prevention of loss of soil during construction by stormwater runoff and/or wind erosion, including protecting topsoil by stockpiling for reuse, and prevention of pollution of the air with dust and particulate matter are clearly identified as requirements for this prerequisite.
5. **A, B, F** The requirements of SSc1 Site Selection do not permit land to be developed that is within 100 feet of a wetland or prime farmland or lower than five feet above the 100-year flood plane. Brownfields are encouraged to be developed (see SSc3 Brownfield Redevelopment). Although virgin forest land should not be developed, it is not addressed in SSc1.
6. **B, D, E** Trees should be protected to the dripline. Water features could be categorized as stormwater detention and would require protection 25 feet beyond the water feature.
7. **C** Only a local, state, or federal agency can issue a brownfield confirmation. The USGBC, AIA, ASHRAE, and FDA do not have the authority to issue brownfield confirmations.
8. **A, B, D** Percolation is associated with increased permeability. Perosity is the measure of a porous medium, such as rock or sediment, and does not apply to stormwater.
9. **A** Bicycle parking for 5 percent FTE for commercial buildings and 15 percent for residential buildings is required to be provided in SSc4.2 Alternative Transportation—Bicycle Storage and Changing Rooms.
10. **E** “In situ” means “in place” in brownfield redevelopment; this term refers to treating contamination on site.

LESSON THREE

3

WATER EFFICIENCY

INTRODUCTION

"Water is vital to the survival of everything on the planet and is limited in supply. The Earth might seem like it has abundant water, but in fact, only 1 percent is available for human use. While the population and the demand on freshwater resources are increasing, supply remains constant."[2]

In this chapter, we will look at the prerequisite and credits for the **Water Efficiency** category, summarized below:

WEp1 Water Use Reduction

WEc1 Water Efficient Landscaping ..2–4 Points

WEc2 Innovative Wastewater Technologies ....2 Points

WEc3 Water Use Reduction..............2–4 Points

"Managing water is a growing concern in the United States. Communities across the country are starting to face challenges regarding water supply and water infrastructure."[2]

"Public-supply withdrawals were more than 43 Bgal/d, (one thousand million gallons per day, abbreviated Bgal/d) during 2000. Public-supply withdrawals during 1950 were 14 Bgal/d. During 2000, about 85 percent of the population in the United States obtained drinking water from public suppliers, compared to 62 percent during 1950. Surface water provided 63 percent of the total during 2000, whereas surface water provided 74 percent during 1950."[3]

"Population growth and economic development are driving a steadily increasing demand for new water supplies; global demand for water has more than tripled over the past half-century. Globally, the largest user of fresh water is agriculture, accounting for roughly three quarters of total use. In Africa, this fraction approaches 90 percent. In the U.S., agriculture accounts for 39 percent of fresh water use, the same fraction used for cooling thermal power plants."[1]

"It is important to emphasize again that we can no longer take water resources for granted if the U.S. is to achieve energy security in the years ahead. This is true of other countries as well, and reflects the strong linkage between water and energy, as well as a growing worldwide water security crisis. Water and energy are also critical elements of sustainable development, a major goal of U.S. foreign policy. Without access to both, economic growth and job creation cannot take place and poverty cannot be averted."[1]

Focusing on reducing the quantity of potable water used by buildings, Water Efficiency prerequisites and credits address easily achieved, low-cost strategies with rapid returns, as well as more complex strategies with significantly higher initial investments that can result in long-term water savings. Strategies introduced in this chapter to reduce building water use and its disposal include monitoring water consumption performance, reducing indoor potable water consumption, lowering water consumption to save energy, and developing water efficient landscaping.

WEp1 WATER USE REDUCTION

Intent

The prerequisite Water Use Reduction is a new requirement in LEED® 2009. Its focus is to increase building water use efficiency to reduce the demand on municipal systems to supply potable water and remove wastewater.

Reducing building water use by increasing efficiency of the plumbing fixtures can decrease the demand for municipally-supplied potable water and wastewater infrastructure.

Reducing the quantity of municipally supplied potable water will lower the amount of chemicals used by the water treatment facility and decrease the energy and the associated greenhouse gases used to perform these processes. Decreasing end-user water consumption reduces operating costs of the building, as well as municipal water supply treatment costs.

Strategies to reduce end-user water consumption can include the following:

- Flow restrictors and/or reduced flow aerators for lavatories, sinks, and shower fixtures
- Automatic faucet sensors and metering controls
- Low consumption fixtures, including high-efficiency water closets and urinals

Low-flow faucets may not be appropriate for kitchen sinks and janitor closets, but water use can be reduced for these types of functions by providing options, such as special-use pot fillers, high-efficiency faucets, and foot pedal-operated faucets.

To establish the best water-saving scenario for the project, teams should evaluate potential water-saving technologies and the impacts of those technologies. Teams should also compare the potential fixtures and fittings against the design-case, water-use calculated Federal Energy Policy Act (EPAct) of 1992 and 2005 baseline to determine if the proposed combination of water-saving plumbing fixtures and fittings meet the project's goals and applicable codes, and will fulfill the requirements of this prerequisite.

Requirements

To attain this prerequisite, project teams must demonstrate a minimum aggregate 20 percent less water use than the baseline calculated for the building, excluding irrigation.

The following fixtures, fittings, and appliances should not be included in the water use calculation:

- Landscape irrigation
- Commercial steam cookers
- Commercial dishwashers
- Automatic commercial ice makers
- Commercial (family-sized) clothes washers
- Residential clothes washers
- Standard and compact residential dishwashers

The Baseline Water Use is calculated based on the water consumption of fixtures in a standard building as mandated under the EPAct of 1992 and 2005. Calculating the projected water use for the building design case typically includes fixtures that use less potable water than minimally compliant EPAct fixtures, and might also use gray or recovered nonpotable water for sewage conveyance (flushing). Keep in mind that the calculation is based on the number of building occupants, not the number of fixtures; no water use reduction can be claimed by simply reducing the number of fixtures.

Following are typical bathroom fixtures and their water usage:

Fixture Type	Baseline
Commercial Toilet	1.6 gallons per flush (gpf) 3.5 gpf (blow-out fixtures)
Commercial Urinal	1.0 gpf
Commercial Lavatory Faucet	2.2 gallons per minute (gpm) at 60 pounds per square inch (psi) for hotel guest rooms and hospital patient rooms .5 gpm at 60 psi for all other commercial applications .25 gallons per cycle for metering faucets

EPAct 1992

It is also important to note that the reference guide allows a 20 percent reduction in time of each use for automatic motion-control or metering sensors on lavatory and sink faucets. In other words, instead of 15 seconds in the case of a standard faucet, a motion-control faucet is assumed to be in use for only 12 seconds.

When calculating water usage, building occupants are considered to be full-time equivalent staff/visitors, and guests are considered transients. Building occupants are assumed to be 50 percent male, and 50 percent female. It is assumed that each occupant will urinate twice a day and defecate once a day. They'll also wash their hands at a lavatory three times a day and use a kitchen sink once a day. If there are showers in a building, it is assumed that on any given day, 10 percent of the building occupants will take a five-minute shower. This equation changes slightly if a building uses a nonpotable water source for sewage conveyance. In such cases, the annual use of nonpotable water may be subtracted from the annual flush volume.

The previous example does not use nonpotable water for sewage conveyance or to otherwise substitute for potable water, nor are any waterless, composting-type water closets used in this example. These technologies, however, are being adopted more frequency. Project teams should always consult with local code authorities about any restrictions on the use of waterless fixtures, or the use of nonpotable water inside a building.

To determine the Full Time Equivalency (FTE), the number of occupants must be identified by occupancy type:

- Full-time staff
- Part-time staff
- Transients, students, visitors, and retail customers
- Residents

The FTE calculation is based upon the number of building occupants during a standard 8-hour daily occupancy for a 40-hour week. An 8-hour occupant has an FTE value of 1.0. Part-time occupants have an FTE based on their hours per day, divided by eight.

Buildings with multiple shifts should use the FTE number for all shifts, and the FTE calculation for each shift must be used consistently for all LEED® credits. For consistency across LEED® credits, calculations should use a one-to-one gender ratio unless specifically warranted.

Referenced Standards

The following referenced standards support this prerequisite:

1. Energy Policy Act, EPAct, of 1992, as amended. This policy addresses energy and water use in commercial, institutional, and residential facilities.
2. Energy Policy Act, EPAct, of 2005. This statute became U.S. law in August 2005. This policy grants authority to regulate alternative energy resources on the outer continental shelf.
3. International Association of Plumbing and Mechanical Officials Publication IAPMO/American National Standards Institute UPC 1-2006, Uniform Plumbing Code 2006, Section 402.0, Water-Conserving Fixtures and Fittings. This publication can be downloaded at: *www.iapmo.org*. IAPMO is an ANSI-accredited code regulating the design, construction, installation, materials, location, operation and maintenance, or use of plumbing fixtures.
4. International Code Council, International Plumbing Code 2006, Section 604, Design of Building Water Distribution System. This publication can be downloaded at *www.iccsafe.org*, and defines maximum flow rated and consumption requirements for plumbing fixtures and fittings for public and private lavatories, showerheads, sink faucets, urinals, and water closets.

Related Credits

If the project team is using rainwater harvesting, or increasing greywater use to decrease the potable water demand, they may also be able to attain the following credits:

- SSc6.1 Stormwater Design—Quantity Control
- SSc6.2 Stormwater Design—Quality Control
- WEc1 Water Efficient Landscaping
- WEc2 Innovative Wastewater Technologies
- WEc3 Water Use Reduction

WEc1 WATER EFFICIENT LANDSCAPING

Intent

Typically, the most economical and easiest approach to reduce or eliminate the need for subsurface or potable water for landscape irrigation is to tailor landscaping to address the project site's microclimate. The use of native or adaptive plants provides self-sustaining landscape that requires little or no watering, because water conservation is naturally built in and attracts native wildlife, and require less fertilizer and pesticides. Water efficient landscaping also helps to maintain natural aquifers by reducing the water needed to support the landscape without drawing from natural reserves.

Ground water is one of our most valuable resources. "When a water-bearing rock readily transmits water to wells and springs, it is called an aquifer. Wells can be drilled into the aquifers and water can be pumped out. Precipitation eventually adds water (recharge) into the porous rock of the aquifer. The rate of recharge is not the same for all aquifers; that must be considered when pumping water from a well. Pumping too much water too quickly draws down the water in the aquifer, and eventually causes a well to yield less and less water, and even run dry. In fact, pumping your well too fast can even cause your neighbor's well to run dry if you both are pumping from the same aquifer."[4]

If an irrigation system is used, pop-up sprinkler heads will typically not meet the requirements of this credit. Drip irrigation, bubblers, soakers, or no irrigation, are strategies that are successful to achieve this credit. Hose bibs are not considered permanent irrigation.

Drip irrigation is typically less expensive to install and has a quicker payback than other irrigation systems, but can have a high design cost. Some municipalities offer rebates or other incentives to encourage the use of water-efficient irrigation systems, dedicated water meters, and rain or moisture sensors.

A different irrigation strategy to reduce wastewater discharge from the site is to harvest greywater and reuse it for irrigation. Keeping more water on the site can help to reduce potable water use, recharge natural aquifers, and reduce demand on the municipal infrastructure for potable water supply and wastewater removal.

Strategies to achieve this credit will vary between regions; for example, in some climates, it may be possible to limit or eliminate permanent irrigation. The site's microclimate, soil type typography, drainage, and sun exposure should always be considered when developing the approach for this credit.

Other examples of region-specific strategies include:

- Drought-tolerant plants and xeriscaping are appropriate design strategies for climates that are hot and dry.
- Native, hardy plants that can survive winter conditions are appropriate for cold climates.
- Native plants combined with rain or moisture sensors are appropriate for hot and humid climates.
- Captured rainwater may also be a good choice for hot and humid climates to reduce or eliminate the need for potable water for irrigation.

Requirements

Project teams have two options to attain this credit. Teams choosing to pursue Option 1 can achieve two points; they can achieve four points if they choose to pursue Option 2.

Option 1: Reduce potable water used for irrigation by 50 percent from the midsummer baseline calculation. Teams must be able to attribute the reduction to one of the following:

- Plant species, density, and microclimate factor
- Irrigation efficiency
- Use of captured rainwater
- Use of recycled wastewater
- Use of water treated and conveyed by a public agency specifically for nonpotable purposes

Groundwater seepage that is pumped away from the immediate vicinity of building slabs and foundations may be used for landscape irrigation to meet the intent of this credit. However, the project team must demonstrate that doing so does not affect site stormwater management systems.

Option 2: Do not use potable water for irrigation.

Teams must first meet Option 1 requirements, and then choose one of two compliance paths:

Path 1: Use only captured rainwater, recycled wastewater, recycled greywater, or water treated and conveyed by a public agency specifically for nonpotable uses for irrigation.

Path 2: Install landscaping that does not require permanent irrigation systems. Temporary irrigation systems used for plan establishment are allowed only if removed within one year of installation.

Water use baseline calculations for landscape irrigation are computed for the month of July. The following information is needed:

- K_L—Landscape Coefficient, volume of water lost via evapotranspiration
- k_s—Species Factor, plant species water needs
- k_d—Density Factor, number of plants and total leaf area of landscape. Density Factor has three ranges: high, medium, and low
- k_{mc}—Microclimate Factor, adjustment of environmental conditions specific to the landscape. Higher kmc conditions occur where evaporative potential is greater, which is usually caused by adjacent heat absorbing and reflective surfaces, or exposure to high winds

Related Credits

If the project team uses native or adaptive vegetation to achieve this credit, they may also be able to attain the following credits:

- SSc5.1 Site Development—Protect or Restore Habitat
- SSc5.2 Site Development—Maximize Open Space
- SSc7.2 Heat Island Effect—Roof

WEc2 INNOVATIVE WASTEWATER TECHNOLOGIES

Intent

The intent of this credit is to reduce building potable water demand for water closet and urinal flushing to reduce the demand on local aquifers. The flush-type fixtures addressed in this credit are regulated by the Energy Policy Act of 1992, and subsequent Department of Energy rulings, requirements of the energy Policy Act of 2005, and editions of the Uniform Plumbing Code and International Plumbing Code.

Unlike faucets, flush-type fixtures, including urinals and water closets, do not have to rely on potable water to function.

Methods of sewage conveyance can include high-efficiency and non-water flush fixtures, rainwater collection, and greywater reuse. Using alternative resources to convey sewage reduces the water drawn from local aquifers, decreases water leaving the site and its draw on municipal infrastructure, as well as associated chemical and energy impacts.

Greywater is wastewater from lavatories, showers, bathtubs, washing machines, and sinks that is not used for disposal of hazardous or toxic ingredients, or wastes from food preparation. Greywater can be reused with only simple filtration. Blackwater is wastewater from toilets, urinals, and kitchen sinks, and cannot be reused without significant treatment. Note that some state and local codes consider wastewater from showers and bathtubs to be blackwater.

When evaluating options to use alternative sewage conveyance, teams should consider the local code restrictions and costs of the potential options. Rainwater collection for nonpotable use has fewer code requirements and costs associated with it than collection of greywater for the same purposes. Greywater collection requires dual sanitary pipes for distribution, additional code requirements for filtration, disinfection, and overflow protection resulting in higher energy cost for operation of the process. Greywater reuse can be a good option if the facility is projected to create large amounts of greywater waste.

When considering on-site water collection, the project team should also compare the projected available on-site water supply with the projected demand for a typical year. The decision to include an alternative wastewater conveyance system should be made in the early stages of design, because it involves considerable design time, as well as higher cost than traditional waste conveyance systems.

Requirements

Project teams have the following two options for attaining this credit:

Option 1: Use water-conserving fixtures, or nonpotable water to reduce the use of municipally-provided potable water for building sewage conveyance by a minimum of 50 percent.

Option 2: Treat 50 percent of wastewater on-site to tertiary standards. Treated water must be used or infiltrated on site.

For both options, teams will need to provide the building baseline calculation for sewage conveyance, and the design case showing on-site water treatment, reflecting the total amount of on-site collected/treated water for sewage conveyance and the percentage of sewage conveyance reduction.

Referenced Standards

There are four referenced standards supporting this credit.

1. Energy Policy Act, EPAct, of 1992, as amended. This policy addresses energy and water use in commercial, institutional, and residential facilities.
2. Energy Policy Act, EPAct, of 2005. This statute became U.S. law in August 2005.
3. International Association of Plumbing and Mechanical Officials Publication IAPMO/American National Standards Institute UPC 1—2006, Uniform Plumbing Code 2006, Section 402.0, Water-Conserving Fixtures and Fittings. This publication can be downloaded at: *www.iapmo.org*. IAPMO is an ANSI-accredited code regulating the

design, construction, installation, materials, location, operation and maintenance, or use of plumbing fixtures.

4. International Code Council, International Plumbing Code 2006, Section 604, Design of Building Water Distribution System. This code can be downloaded at *www.iccsafe.org,* and defines maximum flow rates and consumption requirements for plumbing fixtures and fittings for public and private lavatories, showerheads, sink faucets, urinals, and water closets.

Related Credits

Teams pursuing this credit may also attain the following credits:

- SSc6.1 Stormwater Design—Quantity Control
- SSc6.2 Stormwater Design—Quality Control
- WEp1 Water Use Reduction
- WEc1 Water Efficient Landscaping
- WEc3 Water Use Reduction

Exemplary Performance Credit

One innovation credit point is available in Innovation in Design for teams demonstrating a 100 percent reduction in potable water used for sewage conveyance, or by treating wastewater on-site and either reuse or infiltrate 100 percent of wastewater generated.

WEc3 WATER USE REDUCTION

Intent

The intent of this credit is to increase the water use efficiencies established in WEp1 Water Use Reduction. By increasing water use efficiency, teams can earn two to four points, depending on the reduction percentage they achieve.

Requirements

To achieve this credit, teams should apply the same performance standards and principles established in WEp1, but achieve higher thresholds of water-saving efficiencies. The water-saving percentages for each point is as follows:

Percentage Reduction	Points
30 percent	2
35 percent	3
40 percent	4

Exemplary Performance Credit

Teams can earn one Innovation in Design credit for exemplary performance if they demonstrate 45 percent reduction in projected potable water use.

The following four referenced standards support this credit:

1. Energy Policy Act, EPAct, of 1992, as amended. This policy addresses energy and water use in commercial, institutional, and residential facilities.
2. Energy Policy Act, EPAct, of 2005. This statute became U.S. law in August 2005.
3. International Association of Plumbing and Mechanical Officials Publication IAPMO/ American National Standards Institute UPC 1—2006, Uniform Plumbing Code 2006, Section 402.0, Water-Conserving Fixtures and Fittings. This publication can be downloaded at: *www.iapmo.org*. IAPMO is an ANSI-accredited code regulating the design, construction, installation, materials, location, operation and maintenance, or use of plumbing fixtures.

4. International Code Council, International Plumbing Code 2006, Section 604, Design of Building Water Distribution System. This publication can be downloaded at *www.iccsafe.org*, and defines maximum flow rated and consumption requirements for plumbing fixtures and fittings for public and private lavatories, showerheads, sink faucets, urinals, and water closets.

Related Credits

If the project team is using rainwater harvesting or increasing greywater use to decrease the potable water demand, they may also be able to attain the following credits:

- SSc6.1 Stormwater Design—Quantity Control
- SSc6.2 Stormwater Design—Quality Control
- WEc1 Water Efficient Landscaping
- WEc2 Innovative Wastewater Technologies
- WEc3 Water Use Reduction

ABBREVIATIONS AND ACRONYMS

ANSI	American National Standards Institute
EPA	Environmental Protection Agency
EPAct	Energy Policy Act
FEMA	Federal Emergency Management Act
GPWA	Gross Potable Water Applied (measured in gallons)
IAPMO	International Association of Plumbing and Mechanical Officials
TPWA	Total Potable Water Applied (measured in gallons)
U.S. EPA	United States Environmental Protection Agency

FOOTNOTES

1. Water Industry News, *The Connection: Water Supply and Energy Reserves*, 2009, http://waterindustry.org/Water-Facts/world-water-6.htm.
2. U.S. Environmental Protection Agency, *Why We Need WaterSense*, 2009, http://www.epa.gov/watersense/basic/why_need.htm.
3. U.S. Geological Survey, *Estimated Use of Water in the United States in 2000*, 2005 http://pubs.usgs.gov/circ/2004/circ1268/.
4. U.S. Geological Survey, *Aquifers*, 2009, http://ga.water.usgs.gov/edu/earthgwaquifer.html.

LESSON 3 QUIZ

1. Which three of the following are required to determine the water use baseline for WEc1 Water Efficient Landscaping?

   A. Species Factor

   B. Imperviousness Coefficient

   C. Saturation Factor

   D. Density Factor

   E. Runoff Coefficient

   F. Microclimate Factor

2. What information is needed to calculate the water use baseline for WEc3 Water Use Reduction? (Choose three.)

   A. FTE for building

   B. Annual rainwater collected

   C. Flow rate of fixtures

   D. Greywater use rates

   E. Occupancy

3. Which irrigation type is more efficient?

   A. Sprinkler irrigation

   B. Drip irrigation

4. If choosing Option 1 of WEc1 Water Efficient Landscaping, reduce potable water use by 50 percent, to which four of the following must the water use reduction be attributed?

   A. Irrigation efficiency

   B. Use of captured rainwater

   C. Evapotranspiration

   D. Use of recycled blackwater

   E. Use of recycled greywater

   F. Plant species, density, and microclimate factor

5. In the requirements for WEp1 Water Use Reduction, which four of the following fixtures, fittings, and appliances are excluded from the water use calculation?

   A. Automatic commercial ice makers

   B. Commercial urinals

   C. Standard compact residential dishwashers

   D. Commercial pre-rinse spray valves

   E. Residential dishwashers

   F. Commercial dishwashers

6. What is water from lavatories, showers, and bathtubs that does not contain human waste or toxic substances called?

   A. Greywater

   B. Brownwater

   C. Potable water

   D. Blackwater

   E. Whitewater

7. To what does the term "process water" refer?

   A. Water used to flush urinals

   B. Water used for irrigation

   C. Water used for cooling towers, boilers, and chillers

   D. Water used to supply lavatories and sinks

8. When calculating the baseline water use for WEc1 Water Efficient Landscaping, which month is used in the calculation?

A. April

B. June

C. August

D. July

9. Which three conditions are accounted for in the microclimate factor (k_{mc})?

A. Temperature

B. Average rainfall

C. Wind

D. Slope

E. Region

F. Humidity

10. Which two of the following compliance paths must the project meet to achieve WEc2 Innovative Wastewater Technologies?

A. Treat 50 percent of wastewater on-site to tertiary standards.

B. Treat 50 percent of wastewater off-site to tertiary standards.

C. Reduce building potable water used for sewage conveyance by 50 percent.

D. Reduce building potable water used for sewage conveyance by 20 percent.

QUIZ ANSWERS

Lesson 3

1. **A, D, F** Only the plant species factor, density factor, and microclimate factors are required to determine the landscaping water use baseline.
2. **A, C, E** Annual rainwater collected and greywater are not needed for this calculation.
3. **B** Sprinkler irrigation efficiency is 0.625, but drip irrigation efficiency is 0.90.
4. **A, B, E, F** Evapotranspiration is the loss of water by evaporation from soil and transpiration from plants. Blackwater cannot be used for irrigation.
5. **A, C, E, F** Commercial urinals and commercial pre-rinse spray valves are to be included in the water use reduction calculation.
6. **A** Greywater is wastewater from lavatories, showers, bathtubs, washing machines, and sinks that is not used for disposal of hazardous or toxic ingredients or wastes from food preparation. Greywater can be reused with only simple filtration.
7. **C** Process water is nonpotable water used for industrial processes, including supplying building mechanical equipment.
8. **D** July is the month used for calculating landscaping average water usage during the summer months.
9. **A, C, F** Temperature, wind, and humidity are site-specific conditions. An area that is mostly paved will be hotter and windier than a grassy area with trees.
10. **A, C** See requirements for WEc2 Innovative Wastewater Technologies.

LESSON FOUR

4

ENERGY AND ATMOSPHERE

INTRODUCTION

"On an annual basis, buildings in the United States consume 39 percent of America's energy, and 68 percent of its electricity. Furthermore, buildings generate 38 percent of the carbon dioxide (the primary greenhouse gas associated with climate change), 49 percent of the sulfur dioxide, and 25 percent of the nitrogen oxides found in the air. Currently, the vast majority of this energy is produced from nonrenewable, fossil fuel resources. With America's supply of fossil fuel dwindling, concerns for increasing energy supply security (both for general supply and specific needs of facilities), and the rising impact of greenhouse gases on world climate, it is essential to find ways to reduce load, increase efficiency, and utilize renewable fuel resources in federal facilities."[1]

In this chapter, we will look at the following prerequisites and credits for the **Energy and Atmosphere** category:

EAp1 Fundamental Commissioning of Building Energy Systems

EAp2 Minimum Energy Performance

EAp3 Fundamental Refrigerant Management

EAc1 Optimized Energy Performance.................................. 1–19 Points

EAc2 On-Site Renewable Energy..... 1–7 Points

EAc3 Enhanced Commissioning.........2 Points

EAc4 Enhanced Refrigerant Management...2 Points

EAc5 Measurement and Verification ....3 Points

EAc6 Green Power..............................2 Points

This chapter addresses building sustainability through reduction in building energy use and the use of clean, renewable energy sources. Increased building performance reduces energy demand, resulting in reduced greenhouse gases. Integrating building components using a whole building design approach that includes the building envelope, orientation, water efficiency, heating, ventilation, and air conditioning (HVAC), and lighting systems, will increase building system efficiency, and reduce energy demand and operating costs.

Most of the electricity produced in the United States is generated from fossil fuels, oil, and natural gas. Each source has negative environmental impacts during each step of production, extraction, transportation, refining, and distribution. Nuclear fission produces radioactive waste, which has transportation and disposal issues, and there is also the potential for a catastrophic accident. Hydroelectric generators disrupt natural water flows, and disturb aquatic habitats. Using clean, renewable energy sources, such as wind, solar, and biomass, can reduce our reliance on forms of energy with negative environmental impacts.

Tracking building energy performance includes three components: design, commissioning, and monitoring. The standard for design is based on American Society of Heating, Refrigerating and Air-Conditioning Engineers (ASHRAE) 90.1—2007; this standard establishes building energy performance. If local codes are more stringent, they should be used if approved by the United States Green Building Council (USGBC).

"Building commissioning provides documented confirmation that building systems function according to criteria set forth in the project documents to satisfy the owner's operational needs. Commissioning existing systems may require developing new functional criteria to address the owner's current requirements for system performance."[2]

A building's energy use can be optimized by ongoing measurement and verification. This process provides building owners and managers the information necessary to identify and correct systems that are not functioning optimally. The Measurement and Verification Plan should be based on the Best Management Practices (BMPs) developed by the International Performance Measurement and Verification Protocol (IPMVP). The duration of the plan should extend through at least one year of post-construction occupancy.

Reducing Green House Gases (GHGs) by eliminating chloroflourocarbon (CFC)-based refrigerants in HVAC&R equipment for new buildings is also addressed in this section. Chlorofluorocarbons (CFCs) are part of a larger category of Ozone Depleting Substances (ODS). When released into the atmosphere, ODS destroy ozone molecules in the stratosphere. The Montreal Protocol is a treaty that includes a timetable for the phase out of production and use of ODS.

Using clean, renewable energy sources, such as solar, wind, and biomass instead of traditional energy sources, eliminates the negative environmental consequences associated with energy production. Reducing the use of coal, fossil fuel, natural gas, or nuclear power also reduces the pollutants released during energy production, including sulfur dioxide, nitrogen oxide, and carbon dioxide, which contribute to acid rain, smog, and global warming.

EAp1 FUNDAMENTAL COMMISSIONING OF BUILDING ENERGY SYSTEMS

Intent

This prerequisite requires commissioning for the systems that are essential to the operation of a building. Verification that mechanical and passive heating, ventilation, air conditioning, and refrigeration (HVAC&R) systems, lighting and daylighting controls, and domestic water heating and renewable wind and solar systems are installed and function as designed, will decrease operating and energy costs, and improve productivity of the building occupants.

Minimum energy-related systems to be commissioned are:

- HVAC&R mechanical and passive systems and their controls
- Controls for lighting and daylighting
- Domestic water heating systems
- Renewable energy systems, including wind and solar

Requirements

The following requirements to be completed by the Commissioning Team include:

1. Individual to be designated as Commissioning Authority (CxA). The designated CxA leeds, reviews, and oversees commissioning activities.
   a. CxA to have experience with at least two building projects.
   b. CxA to be independent of the design and construction teams, but can be employed by the same companies providing those services. The CxA can also be an employee or consultant to the owner, but they must meet the CxA qualifications.
   c. All report results, recommendations, and findings to be reported by the CxA directly to the owner.
   d. The CxA can be a member of the design or construction team if the individual has the required experience, and the project is smaller than 50,000 square feet.
2. Owner's Project Requirements (OPR) and Basis of Design (BOD) to be reviewed by CxA. The OPR to be developed by the owner and the BOD to be developed by the design team.
3. Commissioning requirements to be developed and incorporated into construction documents by the CxA.
4. Commissioning plan to be developed and implemented by the CxA.
5. Systems to be commissioned to have their installation and performance verified by the CxA.
6. Summary commissioning report to be completed by the CxA.

The commissioning process is led by the CxA; it is their responsibility to verify that all components of the process are accomplished.

The owner is responsible for establishing their requirements for the project through a document called the OPR. This document should identify the function of the building, as well as any expansion, quality, construction, and operations cost expectations. It should also include sustainability and energy efficiency goals, indoor environmental quality requirements, and expectations for the equipment and systems selected for the building. The document should also describe how the building will be operated, and the level of training needed for the building occupants to understand how the building functions are designed and expected to operate.

The Basis of Design (BOD) is a document developed by the design team to address the primary design assumptions, applicable codes, guidelines and regulations, and performance criteria for the lighting, HVAC&R, hot water, and on-site power systems.

The CxA is responsible to review the OPR and BOD for completeness and accuracy. The CxA is also responsible for verifying that the BOD reflects the owner's requirements. Each entity is always responsible for updating their respective documents. The CxA is then responsible for incorporating this information into the construction documents.

Each commissioning plan is project-specific, but should include the following:

- Overview of commission program
    - Goals and objectives
    - General project information
- Identification of systems to be commissioned
- Identification of the commissioning team
    - Members
    - Roles
    - Responsibilities
- Protocols
    - Communication
    - Coordination
    - Meetings
    - Management
- Commissioning activities
    - Documentation of owner's project requirements
    - Basis of design preparation
    - Procedures for systems functional test
    - Verification of systems performance
    - Deficiencies and resolution reports
    - Acceptance of building systems
- Commissioning milestones

EAp2 MINIMUM ENERGY PERFORMANCE

Intent

The intent of this prerequisite is to reduce or eliminate excessive energy use by achieving minimum energy efficiency levels for building and systems design. Increased energy efficiency reduces a building's demand for and reliance on fossil fuels, oil, natural gas, and other sources of energy that have negative environmental impacts from production, extraction, transportation, refining, and distribution processes.

Hydroelectric generators disrupt natural water flows and disturb aquatic habitats. Reducing the use of coal, fossil fuel, natural gas, or nuclear power also reduces the pollutants released during energy production, including sulfur dioxide, nitrogen oxide, and carbon dioxide, which contribute to acid rain, smog, and global warming.

Optimizing the energy performance of the building will also reduce the costs of operating the building.

Requirements

The requirements for this prerequisite are based on the ASHRAE Standard 90.1—2007, associated ASHRAE guides, and the ENERGY STAR® Program. ASHRAE 90.1—2007 establishes the standard for minimum energy performance of a building. The other referenced standards address energy consumption components, or provide alternative methods to demonstrate compliance with ASHRAE 90.1—2007.

Teams can choose one of the following three options to demonstrate compliance with this prerequisite:

Option 1: Whole Building Simulation

Teams must increase building energy performance by either 10 percent for new buildings, or 5 percent for major renovations to existing buildings, as compared with the building baseline calculation.

Calculations for this option are to use a computer simulation model for the whole building project, according to ANSI/ASHRAE/IESNA Standard 90.1—2007 (without amendments but without addenda) sections 5.4, 6.4, 7.4, 8.4, 9.4, and 10.4.

Option 2: Prescriptive Compliance Path: ASHRAE Advance Energy Design Guide, as it applies to the project scope:

Path 1: ASHRAE Advanced Energy Design Guide for Small Office Buildings 2004.

Buildings must be office occupancy, and less than 20,000 square feet.

Path 2: ASHRAE Advanced Energy Design Guide for Small Retail Buildings 2006.

Buildings must be retail occupancy, and less than 20,000 square feet.

Path 3: ASHRAE Advanced Energy Design Guide for Small Warehouse and Self-Storage Buildings 2008.

Buildings must be warehouse or self-storage occupancy, and less than 50,000 square feet.

Option 3: Prescriptive Compliance Path: Advanced Buildings™ Core Performance™ Guide

Buildings must be less than 100,000 square feet and comply with Section 1: Design Process Strategies, and Section 2: Core Performance Requirements

OR

Buildings must be less than 100,000 square feet, be office, school, public assembly, and retail occupancy, and comply with Sections 1 and 2 of the Core Performance Guide

OR

Buildings must be less than 100,000 square feet and be project types other than health care, warehouse, and laboratory occupancies, and must implement the basic requirements of the Core Performance Guide.

Referenced Standards

There are five referenced standards for this prerequisite, each addressing a different aspect of this credit's requirements.

1. ANSI/ASHRAE/IESNA Standard 90.1—2007: Energy Standard for Buildings Except Low-Rise Residential.

Download this standard at *www.ashrae.org*, which establishes the minimum energy performance standard for building design.

2. ASHRAE Advanced Energy Design Guide for Small Office Buildings 2004. Download this standard at *www.ashrae.org*.
3. ASHRAE Advanced Buildings™ Core Performance™ Guide. Download this standard at *www.advancedbuildings.net*.
4. ENERGY STAR® Program, Target Finder Rating Tool. Download this standard at *www.energystar.gov/index.cfm?=new_bldg_design.bus_target_finder*. ENERGY STAR® Target Finder are tools to help architects and building owners establish aggressive, yet realistic energy targets, and rate the estimated energy use of a building's design.

Related Credits

In addition to the increased energy performance described in this prerequisite, the building envelope, roof, lighting, and HVAC&R systems design can also contribute to building energy savings, which are addressed in the following credits:

- EAc1 Optimize Energy Performance
- SSc7.2 Heat Island Effect—Roof
- SSc8 Light Pollution Reduction

Renewable energy can also contribute to reduced energy demand, which is addressed in the following credits:

- EAc2 On-site Renewable Energy
- EAc6 Green Power

Reducing water use and the energy used to heat domestic water can also contribute to reduced energy demand, which is addressed in WEc3 Water Use Reduction.

EAp3 FUNDAMENTAL REFRIGERANT MANAGEMENT

Intent

This prerequisite focuses on minimizing contributions to the depletion of the stratospheric ozone from refrigerants by reducing GHGs with the elimination of CFC-based refrigerants in HVAC&R equipment for new buildings.

The Montreal Protocol is an international treaty that governs stratospheric ozone protection and research, and the production and use of ozone-depleting substances. It supplies the means of support to end production of ozone-depleting substances, including CFCs. It also provides resources to developing nations to promote the transition to ozone-safe technologies. It requires refrigerants with nonzero operating pressure differential (OPD), such as CFCs and HCFCs, to be phased out by 2030 in developed countries. In compliance with this treaty, CFC production in the United States ended in 1995. In addition, the U.S. Environmental Protection Agency (EPA) has also established regulations directing the responsible management of Ozone Depleting Substances (ODS).

CFCs are part of a larger category of ODS. When released into the atmosphere, ODS destroy ozone molecules in the stratosphere. Although it is standard practice that CFC-based refrigerants are not installed in new base building HVAC&R systems, existing buildings may have CFC-based refrigeration equipment. When reusing existing base building HVAC&R equipment, project teams should complete a comprehensive CFC phase-out conversion.

When selecting refrigeration equipment, project teams should choose refrigerants that have short environmental lifetimes, and should consider Ozone Depletion Potential (ODP), and Global Warming Potential (GWP). They should also consider the phase-out period for CFC substitutes; some refrigerants that are currently available today have short phase-out deadlines. Or, teams should specify HVAC&R equipment that does not use CFC refrigerants. The following table shows examples of refrigerants and their Global Warming Potential (GWP).

Refrigerant	100-year GWP	20-year GWP
R717 (Ammonia)	0	0
R744 (CO_2)	1	1
R290 (Propane)	<20	<20
R404a	3,862	5,651
R134a	1,410	3,590
R410a	2,060	4,095
R407c	1,749	3,869
R417a	2,312	4,577
R22	1,780	4,850
R12	10,720	10,340

www.r744.com/services/files/alternative_refrigerants_ part_1. pdf[4]

Refrigerant leakage is another concern. During installation, operation, charging, servicing, or decommissioning of equipment, refrigerant can leak without detection. To address this issue, the U.S. EPA Clean Air Act of 1990, Section 608 states:

"One 100-pound cylinder of R404a leaked to the atmosphere equals:

- 27 Chevy Suburbans driving 12,000 miles each, or
- 28 acres of forest."[4]

"Require service practices that maximize recovery and recycling of ozone-depleting substances (both chlorofluorocarbons [CFCs] and hydrochlorofluorocarbons [HCFCs] and their blends) during the servicing and disposal of air-conditioning and refrigeration equipment. Set certification requirements for refrigerant recycling and recovery equipment, technicians, and refrigerant reclaimers. Restrict the sale of refrigerant to certified technicians."[3]

"Require persons servicing or disposing of air-conditioning and refrigeration equipment to certify to EPA that they have acquired refrigerant recovery and/or recycling equipment, and are complying with the requirements of the rule."[3]

"Require the repair of substantial leaks in air-conditioning and refrigeration equipment with a refrigerant charge greater than 50 pounds."[3]

Establish safe disposal requirements to ensure removal of refrigerants from goods that enter the waste stream with the charge intact (for example, motor vehicle air conditioners, home refrigerators, and room air conditioners)."[3]

Requirements

This prerequisite requires that project teams do not use CFC-based refrigerants in new base building HVAC&R systems, as well as complete a comprehensive CFC phase-out conversion when reusing existing base building HVAC&R equipment.

Referenced Standard

The referenced standard for this prerequisite is the U.S. EPA Clean Air Act, Title VI, Section 608, Compliance with the Section 608 Refrigerant Recycling Rule. This document can be downloaded at *www.epa.gov/azone/title6/608/608fact.html*, and contains regulations for using and recycling ODS.

EAc1 OPTIMIZED ENERGY PERFORMANCE

Intent

This credit builds on the performance standard established in EAp2 Minimum Energy Performance by further reducing negative environmental impacts related to excessive energy use by achieving increasing levels of energy performance. Increased energy efficiency reduces a building's demand for, and reliance on, fossil fuels, oil, natural gas, and other sources of energy that have negative environmental impacts from production, extraction, transportation, refining, and distribution processes. Hydroelectric generators disrupt natural water flows and disturb aquatic habitats. Reducing the use of coal, fossil fuel, natural gas, or nuclear power also reduces the pollutants released during energy production, including sulfur dioxide, nitrogen oxide, and carbon dioxide, which contribute to acid rain, smog, and global warming.

Requirements

The requirements for this prerequisite are also based on the ASHRAE Standard 90.1—2007, associated ASHRAE guides, and the ENERGY STAR® Program. ASHRAE 90.1—2007 establishes the standard for minimum energy performance of a building. The other referenced standards address energy consumption components, or provide alternative methods to demonstrate compliance with ASHRAE 90.1—2007.

Teams can choose one of the following three options to demonstrate compliance with this credit:

Option 1: Whole Building Simulation

By choosing this compliance option, teams can earn between one and 19 points depending on the level of increased building energy performance achieved as compared with the building baseline calculation.

Calculations for this option are developed by using a computer simulation model for the whole building project, according to ANSI/ASHRAE/IESNA Standard 90.1—2007 (without amendments but without addenda) sections 5.4, 6.4, 7.4, 8.4, 9.4, and 10.4.

Teams can earn points for increased energy performance according to the following chart:

New Buildings	Points
12 percent	1
14 percent	2
16 percent	3
18 percent	4
20 percent	5
22 percent	6
24 percent	7
26 percent	8
28 percent	9
30 percent	10
32 percent	11
34 percent	12
36 percent	13
38 percent	14
40 percent	15
42 percent	16
44 percent	17
46 percent	18
48 percent	19

Option 2: Teams can earn one point for choosing the Prescriptive Compliance Path: ASHRAE Advance Energy Design Guide, as it applies to the project scope:

Path 1: ASHRAE Advanced Energy Design Guide for Small Office Buildings 2004.

Buildings must be office occupancy, and less than 20,000 square feet.

Path 2: ASHRAE Advanced Energy Design Guide for Small Retail Buildings 2006.

Buildings must be retail occupancy, and less than 20,000 square feet.

Path 3: ASHRAE Advanced Energy Design Guide for Small Warehouse and Self-Storage Buildings 2008.

Buildings must be warehouse or self-storage occupancy, and less than 50,000 square feet.

Option 3: Teams can earn between one and three points for choosing the Prescriptive Compliance Path: Advanced Buildings™ Core Performance™ Guide.

Buildings must be less than 100,000 square feet and comply with Section 1: Design Process Strategies and Section 2: Core Performance Requirements. Health care, warehouse, and laboratory occupancies are not eligible for this compliance path.

Path 1: Buildings that are less than 100,000 square feet, are office, school, public assembly, and retail occupancy, and comply with Sections 1 and 2 of the Core Performance Guide, can earn one point.

Options 1 and 2 Topics

1.1 Identifying design intent
1.2 Communicating design intent
1.3 Building configuration
1.4 Mechanical system design
1.5 Construction certification (acceptance testing)
1.6 Operator training and documentation
1.7 Performance data review
2.1 Energy code requirements
2.2 Air barrier performance
2.3 Minimum indoor air quality performance
2.4 Below-grade exterior insulation
2.5 Opaque envelope performance
2.6 Fenestration performance
2.7 Lighting controls
2.8 Lighting power density
2.9. Mechanical equipment efficiency requirements
2.10 Dedicated mechanical systems
2.11 Demand control ventilation
2.12 Domestic hot water system efficiency
2.13 Fundamental economizer performance

Path 2: Teams can earn an additional two points if they implement performance strategies listed in Section 3: Enhanced Performance. One point is available for every three strategies that are implemented.

Option 3 Strategies

3.1 Cool roofs
3.2 Daylighting and controls
3.3 Additional lighting power reduction
3.4 Plug loads, appliance efficiency
3.5 Supply air temperature reset (VAV)
3.6 Indirect evaporative cooling
3.7 Heat recovery
3.8 Night venting
3.9 Premium economizer performance
3.10 Variable speed control
3.11 Demand-responsive buildings (peak power reduction)
3.12 On-site supply of renewable energy
3.13 Additional commissioning
3.14 Fault detection and diagnostics

The following three strategies are not eligible for additional points under this credit:

3.1 Cool roofs
3.8 Night venting
3.13 Additional commissioning

Referenced Standards

There are five referenced standards for this credit, each addressing a different aspect of this credit's requirements.

1. ANSI/ASHRAE/IESNA Standard 90.1—2007: Energy Standard for Buildings Except Low-Rise Residential. This standard establishes the minimum energy performance standard for building design. Download this standard at *www.ashrae.org*
2. ASHRAE Advanced Energy Design Guide for Small Office Buildings 2004. Download this standard at *www.ashrae.org*
3. New Buildings Institute Advanced Buildings™ Core Performance™ Guide. Download this standard at *www.advancedbuildings.net*
4. ASHRAE Advanced Energy Design Guide for Retail Buildings 2006. Download this standard at *www.ashrae.org*
5. ASHRAE Advanced Energy Design Guide Small Warehouses and Self Storage Buildings 2008. Download this standard at *www.ashrae.org*

Exemplary Performance

One innovation credit point is available in Innovation in Design for teams using the Whole Building Simulation Method and demonstrating increased building energy performance of 50 percent for new buildings, and 46 percent for existing buildings as compared with the building baseline calculation.

Related Credits

In addition to the increased energy performance described in this prerequisite, the building envelope, roof, lighting, and HVAC&R systems design can also contribute to building energy savings, which are addressed in the following credits:

- EAp2 Minimum Energy Performance
- SSc7.2 Heat Island Effect—Roof
- SSc8 Light Pollution Reduction

Renewable energy can also contribute to the reduction in energy demand, which is addressed in the following credits:

- EAc2 On-site Renewable Energy
- EAc6 Green Power

Reducing water use and the energy used to heat domestic water can also contribute to reduction in energy demand, which is addressed in WEc3 Water Use Reduction.

EAc2 ON-SITE RENEWABLE ENERGY

Intent

This credit focuses on reducing environmental and economic impacts associated with fossil fuel energy use by recognizing the achievement of increasing levels of on-site renewable energy. Renewable systems include solar, wind, biomass, and biogas energy. Systems eligible for this credit produce electrical power or thermal energy. Passive solar and ground source heat pumps and daylighting do not qualify for this credit because they do not generate power.

Using renewable energy will change the energy performance of the building and require additional commissioning, as well as measurement and verification.

Requirements

Teams can earn between one and seven points for increasing levels of on-site renewable energy. To determine the percentage of renewable energy achieved, teams should use the annual energy cost calculated for EAc1 Optimize Energy Performance, or the U.S. Department of Energy's Commercial Buildings Energy Consumption Survey to determine the estimated energy use. The on-site energy produced by renewable systems should be expressed as a percentage of the annual energy cost for the building.

The thresholds for this credit are as follows:

Percentage Renewable Energy	Points
1 percent	1
3 percent	2
5 percent	3
7 percent	4
9 percent	5
11 percent	6
13 percent	7

Referenced Standard

The referenced standard for this credit is ANSI/ ASHRAE/IESNA Standard 90.1—2007: Energy Standard for Buildings Except Low-Rise Residential. This standard establishes the minimum energy performance standard for building design. Download this standard at *www.ashrae.org*.

Exemplary Performance

One innovation credit point is available in Innovation in Design for teams demonstrating that on-site renewable energy accounts for 15 percent or more of the annual building energy cost.

Related Credits

Renewable systems typically do not contribute to other credits. Prerequisites and credits that benefit On-Site Renewable Energy include:

- EAp1 Fundamental Commissioning of Building Energy Systems
- EAp2 Minimum Energy Performance
- EAc1 Optimize Energy Performance
- EAc5 Measurement and Verification
- EAc6 Green Power

EAc3 ENHANCED COMMISSIONING

Intent

This credit builds on the performance thresholds established in the prerequisite EAp1 Fundamental Commissioning of Building Energy Systems. In addition to the systems commissioned in the prerequisite, this credit expands the systems to be commissioned and the commissioning activities beyond systems performance verification. The scope of additional commissioning should be based on the OPR, and may include commissioning of the following, at the owners discretion:

- Building Envelope
- Stormwater Management Systems
- Water Treatment Systems
- Information Technology Systems

Requirements

This credit requires that, in addition to achieving the EAp1 Fundamental Commissioning of Building Energy Systems, the team implement, or have a contract to implement, the following additional commissioning activities:

1. Prior to the start of the construction documents phase, designate an independent Commissioning Authority (CxA) to lead, review, and oversee the completion of commissioning activities. They must also comply with the following requirements:
   a. CxA must have experience with at least two building projects.
   b. CxA must be independent of the design and construction teams.
   c. CxA cannot be an employee of the design firm, but can be contracted through them.
   d. CxA cannot be an employee of, or contracted through, a contractor

   or construction manager holding construction contracts.
   e. If qualified, the CxA can be either an employee or consultant of the owner.
2. All report results, recommendations, and findings must be reported by the CxA directly to the owner.
3. Independent commissioning authority must conduct one Commissioning Design review of the OPR, BOD, and design documents prior to the mid-construction document phase, and prior to each design submission. All review comments must be rechecked at each review.
4. Simultaneous with A/E review, the independent commissioning authority to review the contractor submittals relative to systems being commissioned to verify OPR and BOD design requirements. CxA must submit the results of this review to the design team and the owner.
5. Commissioning authority must provide owner with a systems manual that contains the information required for recommissioning building systems.
6. Training requirements of building occupants and operating staff must be verified.
7. Commissioning authority must have a contract in place to conduct a review ten months after substantial completion for evaluation of building operation with O&M staff, including a plan for resolution of outstanding commissioning-related issues.

Exemplary Performance

One innovation credit point is available in Innovation in Design for teams that conduct comprehensive envelope commissioning.

Related Credits

Many of the prerequisites and credits include energy-using systems that can benefit from commissioning. Teams should consider adding commissioning requirements to the following prerequisites and credits:

- SSc8 Light Pollution Reduction
- WEc1 Water Efficient Landscaping
- WEc2 Innovative Wastewater Technologies
- WEc3 Water Use Reduction
- EAc1 Optimized Energy Performance
- EAc2 On-Site Renewable Energy
- EAc5 Measurement and Verification
- IEQp1 Minimum IAQ Performance
- IEQc1 Outdoor Air Delivery Monitoring
- IEQc2 Increased Ventilation
- IEQc5 Indoor Chemical and Pollutant Source Control
- IEQc6 Controllability of Systems
- IEQc7 Thermal Comfort

EAc4 ENHANCED REFRIGERANT MANAGEMENT

Intent

This credit builds on the refrigerant management threshold established in EAp3 Fundamental Refrigerant Management to minimize contributions to the depletion of the stratospheric ozone from refrigerants by reducing GHGs with the elimination of CFC-based refrigerants in base building HVAC&R equipment.

Minimizing negative environmental impacts associated with refrigerant use, such as ozone depletion and climate change, should be addressed by the project team through the following objectives:

- Design buildings that do not rely on chemical refrigerants
- Design HVAC&R energy-efficient equipment
- Select refrigerants with zero or low ODP, and minimal direct GWP
- Maintain HVAC&R equipment to reduce refrigerant leakage into the environment

Greenhouse gases and ozone depletion are not a result of the use of refrigerants, but occur when refrigerants are released into the atmosphere, which often occurs from equipment leaks. Natural refrigerants, such as water, carbon dioxide, and ammonia, have a lower potential to damage the atmosphere than manufactured chemical refrigerants.

Requirements

To achieve this credit, teams can earn two points by following one of two compliance paths. The first path is simply to not use refrigerants. The second path requires the team to comply with the following formula, which establishes the maximum threshold for the combined contributions to ozone depletion and global warming potential for the refrigerants, heating, ventilation, air conditioning, and refrigeration used.

Refrigerant Atmospheric Impact = Σ LCGWP + LCODP x $10^5 \leq 100$

To determine the Refrigerant Atmospheric Impact, the Life-Cycle Global Warming Potential and Life-Cycle Ozone Depletion Potential must first be determined using the following formulas:

Life-Cycle Global Warming Potential, LCGWP

$$= \frac{\text{GWPr} \times (\text{Lr} \times \text{Life} + \text{Mr}) \times \text{Rc}}{\text{Life}}$$

Life-Cycle Ozone Depletion Potential, LCODP

$$= \frac{\text{ODPr} \times (\text{Lr} \times \text{Life} + \text{Mr}) \times \text{Rc}}{\text{Life}}$$

The following acronyms and measures are used in the above calculations:

- LCODP: Life-Cycle Ozone Depletion Potential (lb CFC 11/ton-year)
- LCGWP: Life-Cycle Global Warming Potential (lb CO_2/ton-year)
- GWPr: Global Warming Potential of Refrigerant (0 to 12,000 lb CO_2/lbr)
- ODPr: Ozone Depletion Potential of Refrigerant (0 to 0.2 lb CFC 11/lbr)
- Lr: Refrigerant Leakage Rate (0.5 percent to 2.0 percent; default of 2 percent unless otherwise demonstrated)
- Mr: End-of-Life Refrigerant Loss (2 percent to 10 percent; default of 10 percent unless otherwise demonstrated)
- Rc: Refrigerant Charge (0.5 to 5.0 lbs of refrigerant per ton of gross ARI-rated cooling capacity)
- Life: Equipment Life (10 years; default based on equipment type unless otherwise demonstrated)

Teams should use the following formula if multiple types of equipment are used. The formula will provide a weighted average of all base building HVAC&R equipment.

Average Refrigerant Atmospheric Impact =

$$\frac{\sum (\text{LCGWP} + \text{LCODP} \times 10^5) \times Q_{unit}}{Q_{total}} \leq 100$$

The following acronyms are used in the previous calculation:

Q_{unit} = Cooling capacity of an individual HVAC&R or refrigeration unit (tons)

Q_{total} = Total cooling capacity of all HVAC&R or refrigeration

Projects are not allowed to claim zero leakage for the life of the HVAC&R equipment.

For each base building HVAC&R unit, the following pieces of information are required:

- Refrigerant charge (Rc)
- Refrigerant type, used to determine ODP and GWP
- Equipment type, used to determine Life

The 2007 ASHRAE Application Handbook should be consulted to determine Life of equipment. Following are examples of equipment Life:

- Window AC units and heat pumps—10 years
- Unitary, split-packaged AC units and heat pumps—15 years
- Reciprocating compressors, scroll compressors and reciprocating chillers—20 years
- Absorption chillers—23 years
- Water-cooled packaged air-conditioners—24 years
- Centrifugal chillers—25 years

Fire suppression systems that contain ozone-depleting substances, such as CFCs, HCFCs or halons, should not be used to achieve this credit.

Equipment with less than 0.5 pounds of refrigerant, such as small HVAC&R units, standard refrigerators, and small water coolers, are not considered part of the base building system and are not included in the requirements for this credit.

Related Credits

Overall energy performance can be affected by the HVAC&R equipment selected. The following credits address energy use, thermal comfort and refrigerant management, and should be considered with respect to refrigerant selection.

- EAp3 Fundamental Refrigerant Management
- EAp2 Minimum Energy Performance
- EAc1 Optimize Energy Performance
- IEQ 7.1 Thermal Comfort—Design
- IEQ 7.2 Thermal Comfort—Verification

EAc5 MEASUREMENT AND VERIFICATION

Intent

Ongoing accountability for building energy consumption performance minimizes the economic and environmental impacts associated with energy-using systems. This process provides building owners and managers the information necessary to identify and correct systems that are not functioning optimally. Even minor performance modifications can result in substantial savings over the course of a building's lifetime, but can go unrealized without oversight.

Requirements

Teams have two compliance paths to achieve this credit. Using the International Performance Measurement and Verification Protocol, teams can choose Option B from the protocol called the Energy Conservation Measure Isolation, or

Option D, which is a Calibrated Simulation. Energy Conservation Measures (ECM) refers to the installation of equipment or systems, or modifications of equipment or systems for the purpose of reducing energy use and/or cost.

Option B, Energy Conservation Measure Isolation, is a good choice for small and/or simple building projects. The main energy systems are isolated and individually evaluated. The energy savings can be determined by entering the metered data into a simple spreadsheet. This will allow identification of deficiencies and operation errors. Typically, this option is less expensive to operate because, once the meters are installed and tracking is in place, maintaining the system requires little effort.

Option B should be used for projects with the following conditions:

- The interaction of the building's energy conservation measures with other building equipment can be easily measured, or is believed to be insignificant.
- The parameters that affect energy use are simple to monitor.
- If measurement is limited to a few parameters, this option is less complicated and expensive than Option D.
- Meters can serve a dual purpose; for example, submetering is used for both operational feedback and tenant billing.
- Projected baseline energy use can be readily and reliably calculated.

Option D, Calibrated Simulation (Saving Estimation method 2), should be used for buildings with a large number of energy conservation measures, or interacting systems. This option compares the actual energy use of the building and its systems with the performance predicted by a calibrated computer model. The baseline is determined by removing the energy conservation measures from the computer model. Energy savings can be determined by subtracting the baseline energy simulation from the actual energy use.

Option D should be used for projects with the following conditions:

- Projects that have large or complex energy conservation measures that affect building energy use and could benefit from calibration of the as-built energy simulation model.
- Projects for which design could be improved with the information provided from calibration of the as-built energy simulation model, and the breakdown of energy use.
- Projects that could benefit from a breakdown of energy end uses to help determine the most effective areas for energy conservation, such as electrical lighting versus gas water heating.
- Projects that could benefit from the comparison of the calibrated as-built model with the calibrated baseline can show the payback of the capital costs of multiple, interactive conservation measures.

The Measurement and Verification Plan should be based on the Best Management Practices (BMPs) developed by the International Performance Measurement and Verification Protocol. The Measurement and Verification (M&V) plan for both options noted above must include at least one year of post-occupancy evaluation, as well as a process for corrective action if the M&V plan results indicate that energy savings are not being achieved. It should also include the baseline, metering requirements and methodologies for implementing the plan and identify responsibility for the design, coordination, and implementation of the M&V plan. This responsibility should be designated to a member of the design team. Typically, the team member responsible for the energy engineering analysis is also best qualified for this responsibility. Third party verification is also acceptable.

The eight steps to Creating a Measure and Verification Plan arc:

1. List all measures to be monitored and verified.
2. Define the baseline.
3. Define the basis of design and projected savings.
4. Define the general M&V approach.
5. Prepare a project-specific M&V plan.
6. Verify installation and commissioning of energy conservation measures (ECM) or energy-efficient strategies.
7. Determine savings under actual post-installation conditions.
8. Re-evaluate at appropriate intervals.

The referenced standard for this credit is the International Performance Measurement and Verification Protocol, Volume III, EVO 30000.1—2006, Concepts and Options for Determining Energy Saving in New Construction, effective January 2006.

Related Credits

If renewable energy generation systems are being pursued, a M&V plan of these systems may be applicable. Refer to the following related prerequisites and credits:

- EAp2 Minimum Energy Performance
- EAc1 Optimize Energy Performance
- EAc2 On-site Renewable Energy

If the team's commissioning plan included measurement devices and capabilities to track building performance, it can be used as the basis for the M&V plan. Ongoing commissioning programs are especially beneficial in establishing a M&V plan. Refer to the following related prerequisites and credits:

- EAp1 Fundamental Commissioning of Building Energy Systems
- EAc3 Enhanced Commissioning

EAC6 GREEN POWER

Intent

Support of grid-source, renewable energy technology development on a net-zero pollution basis reduces the use of coal, fossil fuel, natural gas, or nuclear power, and also reduces the pollutants released during energy production, including sulfur dioxide, nitrogen oxide, and carbon dioxide, which contribute to acid rain, smog, and global warming. Hydroelectric generators do not release pollutants, but they do disrupt natural water flows and disturb aquatic habitats. Using clean, renewable energy sources, such as solar, wind, and biomass, instead of traditional energy sources, eliminates the negative environmental consequences associated with energy production.

Requirements

To achieve this credit, teams must commit to at least a 2-year renewable contract to provide at least 35 percent of the building's electricity from renewable sources, as defined by the Center for Resource Solutions' Green-e Energy product certification requirements. Quantity of green power purchased is to be based on the quantity of energy consumed, not cost. Acceptable forms of green power are solar, wind, geothermal, biomass, or low-impact hydro.

Teams have two options for measuring the quantity of energy consumed: they can either determine or estimate the baseline energy use.

To determine the baseline energy use, the team can use the annual electricity consumption quantity that is established in EAc1 Optimize Energy Performance. The project owner can contract with the Green-e power provider for at least 35 percent of that quantity for two years.

To estimate baseline energy use, the team should consult the Department of Energy's Commercial Buildings Energy Consumption Survey database to estimate energy use. Electrical intensity factors for various building types in the United States are provided in the database. The energy intensity for the building type is multiplied by the building square footage to estimate baseline energy use. The project owner can contract with the Green-e power provider for at least 35 percent of that quantity for two years.

Teams may obtain a contract for Green-e power under one of the following scenarios:

- If the state has an open electricity market, the building owner may be able to select a Green-e-certified provider. If this is possible, the owner can secure the contract with the provider.
- If the state has a closed electricity market, the governing utility company may have a Green-e-accredited utility program, in which the owner can commit to a 2-year enrollment period. If the utility does not offer a 2-year enrollment option, the owner will need to submit a letter of commitment that they will stay enrolled in the program for at least two years.
- If Green-e-certified power cannot be purchased through a local utility, Green-e-accredited Renewable Energy Certificates (RECs), can be purchased by the owner or project team. They can purchase a quantity of RECs equal to 35 percent of the predicted annual electricity consumption over a 2-year period. Certificates can be purchased all at once or in contracted increments. RECs, also known as "green-tags," compensate Green-e generators for the difference between production costs and the market rate to sell to the grid. Purchasing RECs will not decrease the cost of electricity purchased from the local utility.

If renewable energy is not Green-e-certified, it must meet both of the following criteria for Green-e Energy Certification:

- The energy source must meet the requirements for renewable resources in the Green-e standard.
- The renewable energy supplier has undergone an independent third-party verification that the Green-e standard has been met. Download this standard at *www.green-e.org*

Referenced Standard

The referenced standard for this credit is the Center for Resource Solutions Green-e Product Certification Requirements. Download this standard at *www.green-e.org*. This program is a voluntary certification and verification program for renewable energy products.

Related Credits

Replacing conventional energy sources with renewable energy sources also reduces energy cost, and can positively impact EAc1 Optimized Energy Performance.

On-site renewable energy systems should be commissioned and may also have an effect on the project's roofing installation; refer to the following prerequisites and credits:

- SSc7.2 Heat Island Reduction—Roof
- EAp1 Fundamental Commissioning of Building Energy Systems
- EAc3 Enhanced Commissioning

ABBREVIATIONS AND ACRONYMS

ASHRAE	American Society of Heating, Refrigeration and Air Conditioning Engineers
BMPs	Best Management Practices
BOD	Basis of Design
CBECS	Commercial Buildings Energy Consumption Survey
C&D	Construction and Demolition
CFCs	Chlorofluorocarbons
CFR	Code of Federal Regulations
CO_2	Carbon Dioxide, also known as R744
CRS	Center for Resource Solutions
DEC	Design Energy Cost
DOE	Department of Energy
ECB	Energy Cost Budget
ECM	Energy Conservation Measures
EPA	Environmental Protection
SNAP	Agency Significant New Alternatives Program
FSC	Forest Stewardship Council
GWP	Global Warming Potential
HC	Hydrocarbon
HCFC	Hydroclorofluorocarbons
HFC	Hydrofluorocarbons
HVAC&R	Heating, Ventilating, Air-Conditioning and Refrigeration
IPCC	International Panel for Climate Change
IPMVP	International Performance Measurements and Verification Protocol
Life	Equipment Life
LCGWP	Life-cycle Global Warming Potential
LCODP	Life-cycle Ozone Depleting Potential
Lr	Refrigerant Leakage Rate
M&V	Measurement and Verification
Mr	End of Life Refrigerant Loss
NH_3	Ammonia, also known as R717
ODP	Ozone Depleting Potential
O&M	Operation and Maintenance Manual
OPR	Owner Project Requirements
Q_{unit}	Cooling capacity of an individual HVAC or refrigeration unit (tons)
Q_{total}	Total cooling capacity of all HVAC or refrigeration
REC	Renewable Energy Certificates
Rc	Refrigerant Charge
TAB	Test and Balance Report

FOOTNOTES

1. National Institute of Building Sciences, Whole Building Design Guide, *Optimize Energy Use*, 2009, http://www.wbdg.org/design/minimize_consumption.php.

2. Building and Commissioning Association, Commissioning, 14 Aug. 2009, http://www.bcxa.org/.

3. U.S. Environmental Protection Agency, *Complying With The Section 608 Refrigerant Recycling Rule*, 2009, http://www.epa.gov/Ozone/title6/608/608fact.html#regreqs.

4. Garland, Ted. *Alternative Refrigerants—Part 1*, 2007, http://www.r744.com/services/files/alternative_refrigerants_part_1.pdf.

LESSON 4 QUIZ

1. Of the following, which are the three Energy and Atmosphere prerequisites?

   A. Minimum Energy Performance
   B. Measurement and Verification
   C. Minimum Indoor Air Quality Performance
   D. Fundamental Commissioning of Building Energy Systems
   E. Fundamental Refrigerant Management
   F. Outdoor Air Delivery Monitoring

2. The EAc1 awards points for increasing levels of energy performance. What is the first threshold that can attain a point, and what is the next percentage increase that can attain a point?

   A. 12 percent, 3 percent
   B. 12 percent, 2 percent
   C. 12 percent, 1 percent
   D. 10 percent, 2 percent
   E. 14 percent, 2 percent
   F. 10 percent, 1 percent

3. Which three of the following systems are required to be commissioned for the Energy and Atmosphere prerequisite Fundamental Commissioning of Building Energy Systems?

   A. Building envelope
   B. Lighting and daylighting controls
   C. Mechanical and passive HVAC&R systems
   D. Stormwater management systems
   E. Domestic hot water systems
   F. Water treatment systems

4. Which three of the following should the OPR include?

   A. Primary design assumptions
   B. Owner and user requirements
   C. Energy efficiency goals
   D. Indoor environmental requirements
   E. Commissioning plan
   F. Environmental and sustainability goals

5. In EAc1 Optimize Energy Performance, what is the percentage of increased energy performance over the building baseline calculation that must be achieved to earn a point for Innovation in Design?

   A. 20 percent
   B. 30 percent
   C. 40 percent
   D. 50 percent
   E. 60 percent

6. In EAc2 On-site Renewable Energy, how many points does a team earn by providing 5 percent of the building energy through on-site renewable energy?

   A. 1 point
   B. 2 points
   C. 3 points
   D. 4 points
   E. 5 points

7. From the list below, what two pieces of information are needed to determine the Refrigerant Atmospheric Impact?

A. LCODP

B. ODP

C. Q_{total}

D. LCGWP

E. GWP

8. What is the referenced standard for EAc5 Measurement and Verification?

A. ASHREA 90.1

B. ENERGY STAR® Program

C. Center for Resource Solutions Green-e Product Certification Requirements

D. IPMVP

E. U.S. EPA Clean Air Act

9. To achieve EAc6 Green Power, what percentage of the building electricity must be purchased from a renewable source and for what period of time?

A. 25 percent, 5 years

B. 35 percent, 5 years

C. 25 percent, 1 year

D. 35 percent, 1 year

E. 25 percent, 2 years

F. 35 percent, 2 years

10. Which three of the following are acceptable forms of renewable energy for EAc6 Green Power?

A. Solar

B. Hydro

C. Wind

D. Geothermal

E. Passive solar

F. Nuclear

Lesson 4

1. **A, D, E** Measurement and Verification is a credit in the Energy and Atmosphere category; Minimum Indoor Air Quality Performance is a prerequisite in this category, and Outdoor Air Delivery Monitoring is a credit in the Indoor Environmental Quality category.

2. **B** Teams can earn one point for achieving a 12 percent increase in performance over the baseline calculation. Additional points are awarded for each 2 percent performance increase over 12 percent.

3. **B, C, E** Building envelope, stormwater management systems and water treatment systems are not required to be commissioned in EAp1 Fundamental Commissioning of Building Energy Systems, but can be commissioned in EAc3 Enhanced Commissioning.

4. **B, D, E, F** It is the owner's responsibility to establish the building function operation cost parameters and the expected level of sustainability. It is the design team's responsibility to establish the project's technical parameters in the BOD.

5. **D** Teams can earn between one and 19 points for increasing building energy performance from 12 percent to 48 percent above the building baseline calculation. One additional point is awarded to teams that demonstrate that they have exceeded the 48 percent threshold by 2 percent, and achieve 50 percent increased energy performance.

6. **C** Teams can earn between one and seven points for providing on-site renewable energy. By providing 5 percent of the building's energy needs, the team will earn three points.

7. **A, D** The Life-Cycle Ozone Depleting Potential, and the Life-Cycle Global Warming Potential of a refrigerant are the two measures needed to determine the Refrigerant Atmospheric Impact.

8. **D** The referenced standard for this credit is the International Performance Measurement and Verification Protocol (IPMVP).

9. **F** 35 percent of the building electricity must be purchased from renewable sources for at least two years.

10. **A, C, E** Only solar, wind, and geothermal are acceptable forms of renewable energy for EAc6. Hydro power must be low impact to qualify, passive solar is not quantifiable as an energy source, and nuclear is not a renewable power source.

LESSON FIVE

5

MATERIALS AND RESOURCES

INTRODUCTION

This category focuses on decisions regarding the selection of sustainable building materials, waste reduction, and recycling and their effects to help protect the environment.

In this chapter, we will look at the following prerequisite and credits for the **Materials and Resources** category.

MRp1 Storage and Collection of Recyclables

MRc1.1 Building Reuse—Maintain Existing Walls, Floors and Roof......................1–3 Points

MRc1.2 Building Reuse—Maintain Interior Nonstructural Elements.......................... 1 Point

MRc2 Construction Waste Management......................................1–2 Points

MRc3 Materials Reuse......................1–2 Points

MRc4 Recycled Content...................1–2 Points

MRc5 Regional Materials................1–2 Points

MRc6 Rapidly Renewable Materials..... 1 Point

MRc7 Certified Wood........................... 1 Point

Providing recycling opportunities for all building occupants allows a significant portion of the solid waste stream to be diverted from landfills, and provides materials for new products that would otherwise be manufactured from virgin materials. Recycling material that contains toxic components, such as computers, paints, and batteries, prevents air and groundwater contamination.

Similar to recycling opportunities for building occupants, recycling construction and demolition debris reduces the demand for virgin materials by recovering existing materials. It also reduces the negative environmental impacts from extraction, processing, and transportation.

Reusing buildings can significantly reduce and redirect waste away from landfills. This practice also has the benefit of causing fewer disturbances to the local habitat, and can reduce or eliminate the need for new utility or transportation infrastructure. Avoiding building demolition costs by reusing major building components can also reduce construction costs and duration.

Sustainable material selection can reduce the negative environmental impact of extraction, transportation, and depletion of finite raw materials, and long-cycle renewable materials. Using materials with recycled content reduces the demand for virgin materials and the production of solid waste volumes. Using regional materials lowers the distance the material must travel from the point of extraction to the site. Choosing rapidly renewable materials reduces the use of raw materials, and materials with long regrowth periods.

Using certified wood encourages environmentally responsible forest management. Irresponsible logging practices results in deforestation, loss of wildlife habitat, soil erosion, stream sedimentation, waste generation, and water and air pollution.

MRp1 STORAGE AND COLLECTION OF RECYCLABLES

Intent

Providing easily accessible recycling bins in appropriate locations so that building end-users can responsibly dispose of their waste and recyclables can significantly divert solid waste from landfills. Recycling solid waste and materials that contain toxic components, such as computers, paints, and batteries, prevents land, water, and air pollution.

Diverting solid waste provides materials for new products that would otherwise be manufactured from virgin materials, which avoids extraction of raw materials and the associated negative environmental impacts.

"Recycling aluminum cans, for example, saves 95 percent of the energy required to make the same amount of aluminum from its virgin source, bauxite."[1]

To make recycling as easy as possible, project teams should review building occupants' recycling needs to provide recycling containers that are of sufficient size, and that are easily accessible. Areas for recycling storage should also be easily accessible.

To achieve this prerequisite, teams must provide an easily accessible, dedicated area for the collection and storage of materials for recycling for the entire building. Materials collected for recycling must include paper, corrugated cardboard, glass, plastics, and metals.

MRc1.1 BUILDING REUSE—MAINTAIN EXISTING WALLS, FLOORS, AND ROOF

Intent

Extending the life-cycle of existing building stock preserves resources, retains cultural resources, and reduces waste while lowering the environmental impacts of new buildings as they relate to materials manufacturing and transport. Reusing buildings can appreciably reduce and redirect waste away from landfills. This practice also has the benefit of causing fewer disturbances to the local habitat, and can reduce or eliminate the need for new utility or transportation infrastructure. Avoiding building demolition costs by reusing major building components can also reduce construction costs and duration.

To determine if an existing building can be reused, teams should develop a comprehensive reuse management plan to evaluate the anticipated materials that can be saved. This exercise will help teams assess the building components that can be reused, and determine if the project meets the requirements of this credit. The comprehensive review should include the following steps:

- Review the existing building's structural shell to determine if it meets the project's needs.
- Determine if the existing building's structure can accommodate the proposed programming and space planning.
- If the structural shell cannot be reused, consider reusing or preserving the building's facade, particularly in urban areas.
- Review the pros and cons of the structure's attributes and potential benefits, including solar, transportation access, existing air quality levels, and opportunity to upgrade outdated building components, such as insulation, or glazing.
- Identify existing hazardous materials, including asbestos and lead paint, so the appropriate removal and/or isolation measures can be applied.

Requirements

To achieve this credit, teams must maintain a percentage of the existing building structure, including structural floor and roof decking, building envelope, including exterior skin, and framing. Window assemblies and nonstructural roofing are not included in this requirement.

If the project includes an addition larger than twice the original size of the building, this credit is not applicable.

If hazardous materials are remediated within the project scope, they must be excluded from the calculation of the percentage maintained.

The following table displays the points available for increasing levels of building reuse attained.

Building Reuse	Points
55 percent	1
75 percent	2
95 percent	3

To determine the amount of the building that is reused, surface areas of major existing structural and envelope elements are quantified by square foot. Structural support elements, such as columns and beams, are not quantified separately, because they are considered part of the larger surfaces they support.

Related Credits

This credit involves selective demolition, reusing materials, and responsible disposal of solid waste, and may help the team in achieving the following credits:

- MRc2 Construction Waste Management
- MRc3 Materials Reuse

MRc1.2 BUILDING REUSE—MAINTAIN INTERIOR NONSTRUCTURAL ELEMENTS

Intent

This credit is very similar to MRc1.1 Building Reuse—Maintain Existing Walls, Floors, and Roof, except it focuses on maintaining interior walls, doors, floor coverings, and ceiling systems. As with the previous credit, the intent is to extend the life-cycle of existing building stock to preserve resources, retain cultural resources, and reduce waste, while lowering the environmental impacts of new buildings as they relate to materials manufacturing and transport.

To determine if an existing building can be reused, teams should develop a comprehensive reuse management plan to evaluate the anticipated materials that can be saved. This exercise will help teams assess the building components that can be reused, and determine if the project meets the requirements of this credit.

If the project includes an addition larger than twice the original size of the building, this credit is not applicable.

Requirements

To achieve this credit, teams need to demonstrate that at least 50 percent (by area) of the interior nonstructural elements are retained in the completed buildings. Nonstructural elements include interior walls, doors, floor coverings, and ceiling systems.

Related Credits

This credit involves selective demolition, reusing materials, and responsible disposal of solid waste, and may help the team achieve the following credits:

- MRc2 Construction Waste Management
- MRc3 Materials Reuse

MRc2 CONSTRUCTION WASTE MANAGEMENT

Intent

Minimizing the factors that produce waste by controlling production at its source is the most effective approach to reducing quantities of waste generated. Redirecting construction and demolition waste from landfills and incinerators can be accomplished by channeling recyclable material back to the manufacturing process, and reusable material to appropriate sites. Recycling construction and demolition waste reduces demand for virgin materials, and reduces the environmental impacts associated with extraction, processing, and transportation.

Requirements

To achieve this credit, teams must recycle and/or salvage nonhazardous construction and demolition debris. To accomplish this objective, teams must develop and implement a construction waste management plan that, at a minimum, identifies the materials to be recycled or salvaged, and whether they will be sorted on-site or comingled. All employees of the general contractor and subcontractors must understand and participate in the construction waste management plan.

Availability of recycling resources and salvaging of construction waste varies by region. Most communities offer metal, vegetation, concrete, and asphalt recycling, but recycling of paper, corrugated cardboard, plastics, and clean wood varies depending on the recycling infrastructure of the area. The condition of the material can also affect its recyclability. Landfill space and tipping fees are other issues that can affect waste disposal availability and cost. Teams may be able to sell salvaged material, such as furniture, computers, whiteboards, lockers, doors, lighting, and plumbing fixtures. Team should explore both options in working to attain this credit.

The following should be identified in the construction waste management plan:

- Establish goals for landfill diversion.
- Designate an area of the site for construction and demolition waste.
- Train workers on recycling protocol.
- Label recycling containers effectively.
- Require monthly feedback and reporting on the waste management plan.

The minimum percentage debris to be recycled or salvaged for each point threshold is as follows:

Recycled or Salvaged	Points
50 percent	1
75 percent	2

The calculation for this credit is based on the amount of waste diverted from the landfill or incineration as compared to the total amount of construction waste generated on-site, excluding hazardous waste, excavated soil, and land-clearing debris. Excavated soil, land-clearing debris, and hazardous waste do not contribute to this credit; however, projects that crush or reuse existing concrete, masonry, or asphalt on-site can include these materials in the calculations for this credit. Calculations can be documented by weight or volume, but this measurement must be consistent throughout project documentation.

Exemplary Performance

One Innovation in Design point for exemplary performance will be awarded to projects that can demonstrate that 95 percent or more of the total construction waste was diverted.

Related Credits

Projects that are reusing existing buildings, but do not meet the requirement for MRc1 Building Reuse, can apply the reused building elements toward this credit.

MRc3 MATERIALS REUSE

Intent

Using salvaged, refurbished, or reused building materials helps reduce demand for virgin materials, and, as a result, reduces waste and the associated environmental impacts from its extraction, processing, and transportation.

Requirements

To achieve this credit, teams must incorporate salvaged, refurbished, or reused building materials into the design of the project. The sum of the materials must comprise at least 5 percent or 10 percent of the total value of materials in the project. Teams achieving 5 percent reused materials will earn one point; teams achieving 10 percent reused materials will earn two points.

To qualify for this credit, salvaged, refurbished, or reused items found on the project site must no longer be able to serve their original function, and must then be installed for a different use, or in a different location. For example, wood flooring may be salvaged from a building and used to create cabinets for the project. Items such as walls, ceilings, and flooring that are reused for their original purpose cannot contribute to this credit, but can contribute to MRc1.2 Building Reuse—Maintain Interior Nonstructural Components. In addition, mechanical, electrical, and plumbing components and appliances are exempt from contributing to this credit.

If the salvaged, refurbished, or reused items are found off-site, the previously described constraints do not apply, with the exception of mechanical, electrical, and plumbing components, and specialty equipment (such as elevators), which are still exempt from contributing to this credit. If furniture is reused from the owner's previous facility, it must be older than two years to qualify for contribution to this credit.

Exemplary Performance

One Innovation in Design point for exemplary performance is available for projects that demonstrate that the value of salvaged or reused materials is at least 15 percent of the total material cost for the building.

Related Credits

It is recommended that projects using salvaged, refurbished, or reused building materials include a comprehensive reuse management plan. This plan will help teams to determine if they meet the requirements for MRc1 Building Reuse, MRc2 Construction Waste Management, and this credit.

Remanufactured materials cannot contribute to this credit, but they can contribute to MRc2 Construction Waste Management, and MRc4 Recycled Materials.

Project material costs used for this credit should be consistent with the following credits:

- MRc4 Recycled Content
- MRc5 Regional Materials
- MRc6 Rapidly Renewable Materials

MRc4 RECYCLED CONTENT

Intent

This credit focuses on incorporating recycled material into the project to reduce the demand for virgin material. There are two types of recycled content: postconsumer and preconsumer. Postconsumer recycled content is comprised of materials that can no longer be used for their original purpose. Preconsumer recycled content consists of raw material diverted from the waste stream during the manufacturing process.

Postconsumer recycled content is of greater impact than preconsumer recycled content that reuses industrial waste. This is because it extends the life-cycle of a material, and reduces the need for raw materials and the environmental impacts associated with its extraction, manufacturing, and transportation.

The availability of products with recycled content is continuing to expand. Many common construction materials, such as steel, gypsum board, and acoustical ceiling tile, contain recycled content due to their manufacturing processes. Other building products with recycled content include concrete, masonry, metals, ceramic tile, rubber flooring, and wall base.

Requirements

To achieve this credit, teams need to demonstrate that materials with recycled content are used for at least 10 percent or 20 percent of the total value of materials in the project. For the purposes of this credit, postconsumer recycled content is figured at 100 percent of its full value, but preconsumer recycled content is figured at 50 percent of its full value. When figuring the recycled content of an assembly, the value is determined by the percent of weight of the recycled content, as compared to the overall weight of the item. Teams achieving 10 percent recycled content will earn one point; teams achieving 20 percent recycled content will earn two points.

Materials that are reclaimed from the same manufacturing processes do not contribute to this credit because, when waste is incorporated back into the same manufacturing process, there is no material diverted from the waste stream. In addition, mechanical, electrical, and plumbing components and appliances are exempt from contributing to this credit.

Furniture can contribute to achieving this credit, but it must be consistently tracked throughout the project.

There are two methods teams can use to calculate the recycled content for the project. They can multiply the total construction cost by 0.45, or the actual materials cost can be totaled using the project's schedule of values with the percentage of recycled content broken out.

Referenced Standard

The referenced standard for this credit is International Standard ISO 14021—1999 —Environmental Labels and Declarations—Self-Declared Environmental Claims (Type II Environmental Labeling). This standard can be downloaded at *www.iso.org*.

Exemplary Performance

One Innovation in Design point for exemplary performance is available for projects that demonstrate achievement of a minimum total value of 30 percent for recycled content.

Related Credits

Coordination of recycled material procurement and the waste management plan can be beneficial to the following credits:

- MRc2 Construction Waste Management
- MRc3 Materials Reuse

Purchasing regionally manufactured products with recycled content can contribute to MRc5 Regional Materials.

Project material costs used for this credit should be consistent with the following credits:

- MRc3 Materials Reuse
- MRc5 Regional Materials
- MRc6 Rapidly Renewable Materials

MRc5 REGIONAL MATERIALS

Intent

The focus of this credit is to support regional manufacturing of building supplies to reduce the environmental impacts associated with their transportation.

Requirements

Teams pursuing this credit must demonstrate that a minimum of 10 percent or 20 percent of building materials or products have been extracted, harvested, or recovered, as well as manufactured, within 500 miles of the project site. Teams demonstrating that at least 10 percent of the materials satisfy the requirements of this credit will earn one point; teams demonstrating that at least 20 percent of the materials satisfy the requirements of this credit will earn two points.

Research for this credit should begin early in the design process to find materials or products that are extracted, harvested, or recovered and manufactured within 500 miles of the project site to determine if the items they have identified for the project are available, and will satisfy the credit's requirements. The qualification of a material or product can require the team to research its components. For example, if the glass for a window comes from Moberly, MO, the aluminum comes from Chicago, IL, and the windows are assembled in Fort Scott, KS, the final assembly is Fort Scott, KS. Additionally, if the components are extracted and manufactured within 500 miles of the project site, but the final assembly is not within the 500-mile radius, the product cannot contribute to the credit.

If building components are assembled on-site, the individual components that are extracted within 500 miles of the site can be counted toward the credit.

Products that contain components that are both from within the 500-mile radius and beyond the 500-mile radius can only count the percentage of components from within the radius.

Furniture and furnishings cannot contribute to this credit. In addition, mechanical, electrical, and plumbing components and appliances are exempt from contributing toward this credit.

Reused and salvaged materials used to meet MRc3 Materials Reuse requirements can also contribute to this credit.

There are two methods that teams can use to calculate the regional materials for the project. They can multiply the total construction cost by 0.45, or the actual materials cost can be totaled using the project's schedule of values, with the percentage of regional materials separated out.

Exemplary Performance

One Innovation in Design point for exemplary performance is available for projects that demonstrate that the value of regionally extracted, harvested, recovered, or manufactured materials, is 30 percent or more of the total materials value of the project.

Related Credits

Specifying regional materials could potentially affect the levels of achievement for the following credits:

- MRc3 Materials Reuse
- MRc4 Recycled Content
- MRc6 Rapidly Renewable Materials

Project material costs used for this credit should be consistent with the following credits:

- MRc3 Materials Reuse
- MRc4 Recycled Content
- MRc6 Rapidly Renewable Materials

MRc6 RAPIDLY RENEWABLE MATERIALS

Intent

The focus of this credit is to reduce the impacts resulting from the extraction and processing of virgin material by increasing demand for building products that incorporate rapidly renewable materials.

Rapidly renewable materials include bamboo, cotton, wool, cork, and jute. Because they can be planted and harvested in a cycle of 10 years or less, they are replenished faster than traditional materials. Using these materials can help reduce the depletion of final raw materials, and long-cycle renewable materials.

Rapidly renewable materials can be incorporated into a project for flooring, insulation, cabinetry, paints, and fabrics. Available products include bamboo flooring, sunflower seed board panels, wheatboard cabinetry, wool carpeting, cork flooring, bio-based paints, geo-textile fabrics made of coir or jute, soy-based insulation, and form-release agent and straw bales.

Requirements

To achieve this credit, teams must demonstrate that they have used rapidly renewable material for at least 2.5 percent of the total value of all building materials and products based on cost.

Furniture and furnishings are excluded from calculations for this credit unless they are consistently tracked across credits MRc3 through MRc7.

Mechanical, electrical, and plumbing components and appliances cannot contribute toward this credit.

There are two methods that teams can use to calculate the rapidly renewable material content for the project. They can multiply the total construction cost by 0.45, or the actual materials cost can be totaled using the project's schedule of values with the percentage of rapidly renewable material content separated out.

If figuring the rapidly renewable material content of an assembly, the value is determined by the percent of weight of the rapidly renewable material content, as compared to the overall weight of the item.

Exemplary Performance

One Innovation in Design point for exemplary performance is available for projects that demonstrate that the content of rapidly renewable materials is 5 percent or more.

Related Credits

If teams are using rapidly renewable resources, such as cork and bamboo, they may find that acquisition of these materials affects their ability to achieve MRc5 Regional Materials, because the source for these materials may be beyond the 500-mile radius.

Project material costs used for this credit should be consistent with the following credits:

- MRc3 Materials Reuse
- MRc4 Recycled Content
- MRc5 Regional Materials

MRc7 CERTIFIED WOOD

Intent

This credit focuses on supporting responsible forest management. Without oversight, irresponsible logging can result in deforestation, loss of wildlife habitat, soil erosion, stream sedimentation, waste generation, and water and air pollution. The Forest Stewardship Council (FSC) provides a seal of approval to forest managers who demonstrate environmentally and socially responsible forest management practices, such as promoting sustainable timber harvesting, preserving wildlife habitat and biodiversity, maintaining soil and water quality by minimizing harmful chemicals, and conserving endangered old-growth forests. FSC supports the rights of indigenous people and following all applicable laws in treaties. It also includes forest workers and forest-dependent communities as stakeholders and beneficiaries of responsible forest management. Responsible forest practices help stabilize economies and preserve forestland for future generations.

Requirements

To attain this credit, teams must demonstrate that a minimum of 50 percent, based on cost, of the wood-based material and products permanently installed in the building are FSC-certified.

Chain of Custody (COC)

All permanently installed wood products are required to be tracked through vendor invoices, including all FSC-certified and non-certified wood products purchased by general contractors and subcontractors.

Each vendor invoice must include the following information for each wood product, which must be presented as an individual line item:

1. FSC- or non-FSC-certified product
2. Dollar value of each product
3. Vendor's COC certification number for FSC products

Vendors submitting invoices for FSC-certified products must be COC-certified by an FSC-accredited certifier.

An exception will be made if separating the wood products into individual line items would result in an exceptionally long document. In this instance, the invoice should indicate the combined value of the wood products, and include the following information:

1. Vendor's COC certification number
2. Letter from the vendor stating that the invoiced products are FSC-certified, and indicating if the products are FSC Pure, FSC Mixed Credit, or FSC Mixed (NN) percent

Only new wood products can contribute to this credit. FSC Recycled and FSC Recycled Credit wood products cannot contribute to this credit, but may be used to contribute to MRc4 Recycled Content. Reclaimed, salvaged, or recycled wood products also cannot contribute. FSC Mixed (NN) percent products can contribute at the indicated percentage.

If FSC-certified wood products are used for temporary assemblies, teams can choose to include the materials in the calculations for this credit. If they do, all wood products used for temporary assemblies must be included. If temporary assemblies are purchased for multiple projects, they can only be applied to one project. Temporary assemblies include formwork, bracing, scaffolding, sidewalk protection, and guardrails.

The referenced standard for this credit is the Forest Stewardship Council's Principles and Criteria. Download this standard at *www.fscus.org*.

Exemplary Performance

One Innovation in Design point for exemplary performance is available for projects that demonstrate that a minimum of 95 percent or more of the project's new wood is FCS-certified.

Related Credits

Teams pursuing this credit may also be able to contribute, or may be influenced by the following credits:

- MRc5 Regional Materials
- IEQ4.4 Low-Emitting Materials—Composite Wood and Agrifiber Products

ABBREVIATIONS AND ACRONYMS

COC Chain of Custody
FSC Forest Stewardship Council

FOOTNOTES

1. U.S. Environmental Protection Agency, *Does the Transportation and Reprocessing of Recyclables Outweigh the Energy Savings?*, 2008, http://www.epa.gov/Region4/waste/rcra/mgtoolkit/FAQs.html.

LESSON 5 QUIZ

1. Which three of the following materials are required to be collected for recycling in MRp1 Storage and Collection of Recyclables?

   A. Wood

   B. Concrete

   C. Masonry

   D. Corrugated cardboard

   E. Plastics

   F. Metal

2. What percentage of the existing walls, floors, and roof must be retained to achieve two points for MRc1.1 Building Reuse—Maintain Existing Walls, Floors, and Roof?

   A. 55 percent

   B. 65 percent

   C. 75 percent

   D. 85 percent

   E. 95 percent

3. What percentage of nonhazardous construction and demolition debris must teams recycle and/or salvage to earn one point for MRc2 Construction Waste Management?

   A. 50 percent

   B. 60 percent

   C. 70 percent

   D. 80 percent

   E. 90 percent

4. How many points will a team earn if 20 percent of the total value of materials in the project contains recycled content?

   A. 5

   B. 2

   C. 3

   D. 6

   E. 1

5. A project team is planning to salvage the wood floors from existing portions of the building and intends to install the flooring in new areas of the building. Will this strategy help the team earn a point for MRc3 Material Reuse?

   A. Yes

   B. No

6. Which three of the following materials are considered rapidly renewable?

   A. Wood

   B. Cotton

   C. Stone

   D. Minerals

   E. Cork

   F. Bamboo

7. To confirm Chain of Custody, which three items must be included on each vendor invoice?

   A. Species of wood

   B. Harvest location of wood

   C. Installed location of wood

   D. FSC- or non-FSC-certified product

   E. Dollar value of each product

   F. Vendor's COC certification number

8. A project team that is pursuing MRc7 Certified Wood is using a strategy of combining new FSC-certified wood with reclaimed wood from a demolished building. What percentage of reclaimed wood will the team need to document to qualify for this credit?

A. 0 percent

B. 15 percent

C. 20 percent

D. 25 percent

E. 30 percent

9. What percentage of the interior nonstructural elements must be retained in the completed building to achieve MRc1.2 Building Reuse—Maintain Interior Nonstructural Elements?

A. 15 percent

B. 20 percent

C. 25 percent

D. 30 percent

E. 50 percent

10. Which three of the following materials cannot be included in the calculations to earn MRc2 Construction Waste Management?

A. Excavated soil

B. Hazardous waste

C. Gypsum board

D. Cardboard

E. Land-clearing debris

F. Windows

QUIZ ANSWERS

Lesson 5

1. **D, E, F** Materials collected for recycling must include paper, corrugated cardboard, glass, plastics, and metals.
2. **C** Teams pursuing MRc1.1 Building Reuse—Maintain Existing Walls, Floors, and Roof can earn up to three points for maintaining existing building components as follows: one point for achieving 55 percent, two points for achieving 75 percent, and three points for achieving 95 percent.
3. **A** Teams pursuing MRc2 Construction Waste Management can earn one point for recycling and/or salvaging 50 percent of nonhazardous construction and demolition debris.
4. **B** Teams achieving 10 percent recycled content will earn one point; teams achieving 20 percent recycled content will earn two points.
5. **B** To qualify for this credit, salvaged, refurbished, or reused items found on the project site must no longer be able to serve their original function, and must then be installed for a different use, or in a different location.
6. **B, E, F** Rapidly renewable materials include bamboo, cotton, and cork. Wood, stone, and minerals cannot be planted and harvested in a cycle of ten years or less.
7. **D, E, F** Each vendor invoice must include the following information, which must be presented as an individual line item, for each wood product:
   1. FSC- or non-FSC-certified product
   2. Dollar value of each product
   3. Vendor's COC certification number
8. **A** Only new wood products can contribute to this credit.
9. **E** To achieve this credit, teams need to demonstrate that at least 50 percent (by area) of the interior nonstructural elements are retained in the completed buildings. Nonstructural elements include interior walls, doors, floor coverings, and ceiling systems.
10. **A, B, E** The calculation for this credit is based on the amount of waste diverted from the landfill or incineration, as compared to the total amount of construction waste generated on-site. It does not include hazardous waste, excavated soil, and land-clearing debris.

LESSON SIX

6 INDOOR ENVIRONMENTAL QUALITY (IEQ)

INTRODUCTION

Building occupant comfort, well-being, and productivity can be affected by the actions taken and decisions made by the design team and the contractor. Teams can improve the indoor environment of a building by

- limiting potential indoor contaminant sources;
- minimizing the introduction of contaminants from potential outdoor sources;
- monitoring the quality of air supplied, limiting exposure to tobacco smoke;
- reducing indoor air quality (IAQ) problems resulting from construction activities and materials; and
- allowing occupants to control lighting and thermal comfort in their work areas and providing daylighting and views.

In this chapter, we will look at the prerequisites and credits for the Indoor Environmental Quality category, summarized below:

IEQp1 Minimum IAQ Performance

IEQp2 Environmental Tobacco Smoke (ETS) Control

IEQc1 Outdoor Air Delivery Monitoring .... 1 Point

IEQc2 Increased Ventilation .... 1 Point

IEQc3.1 Construction IAQ Management Plan—During Construction .... 1 Point

IEQc3.2 Construction IAQ Management Plan—Before Occupancy .... 1 Point

IEQc4.1 Low-Emitting Materials—Adhesives and Sealants .... 1 Point

IEQc4.2 Low-Emitting Materials—Paints and Coatings .... 1 Point

IEQc4.3 Low-Emitting Materials—Flooring Systems .... 1 Point

IEQc4.4 Low-Emitting Materials—Composite Wood and Agrifiber Products................ 1 Point

IEQc5 Indoor Chemical and Pollutant Source Control .... 1 Point

IEQc6.1 Controllability of Systems—Lighting.. 1 Point

IEQc6.2 Controllability of Systems—Thermal Comfort .... 1 Point

IEQc7.1 Thermal Comfort—Design...... 1 Point

IEQc7.2 Thermal Comfort—Verification .... 1 Point

IEQc8.1 Daylight and Views—Daylight.. 1 Point

IEQc8.2 Daylight and Views—Views... 1 Point

"Americans spend about 90 percent of their time indoors, where concentrations of pollutants are often much higher than those outside. Risk assessments performed for radon, environmental tobacco smoke (ETS), and lead have shown that health risks are substantial. Thousands of chemicals and biological pollutants are found indoors, many of which are known to have significant health impacts, both indoors and in other environments."[1]

Design decisions, construction processes, and human activity that contribute to the indoor environmental quality addressed in this chapter include occupant health, safety, and comfort, energy consumption, air change effectiveness, and air contamination management.

IEQp1 MINIMUM INDOOR AIR QUALITY PERFORMANCE

Intent

Buildings with poor ventilation can be uncomfortable, stuffy, foul-smelling, and/or unhealthy for the building occupants. Providing adequate ventilation improves occupant comfort, well-being, and productivity. Indoor air quality (IAQ) can be improved by limiting potential indoor contaminant sources, minimizing the introduction of contaminants from potential outdoor sources, and determining and maintaining at least a minimum zone for outdoor air flow and the minimum outdoor air intake flow required by American Society of Heating, Refrigerating, and Air-Conditioning Engineers (ASHRAE) 62.1—2007.

Requirements

To achieve this credit, teams must follow the appropriate path for the type of ventilation provided.

Path 1: Mechanically Ventilated Spaces

Teams must meet the minimum requirements of ASHRAE 62.1—2007, sections 4 through 7, with errata, but without addenda.

Systems must be designed using the ventilation rate procedure described in the referenced standard or with local code, whichever is more stringent. The code that requires the most outside area is considered the most stringent.

Path 2: Naturally Ventilated Spaces

Teams must comply with ASHRAE 62.1—2007, paragraph 5.1, with errata, but without addenda.

Referenced Standard

The referenced standard for this credit is ASHRAE 62.1—2007: Ventilation for Acceptable Indoor Air Quality. Download this standard at *www.ashrae.org*.

Related Credits

IAQ can be enhanced by Commissioning and Measurement and Verification addressed in the following credits:

- EAp1 Fundamental Commissioning
- EAc3 Enhanced Commissioning
- EAc5 Measurement and Verification

Projects located in dense neighborhoods, near heavy traffic, and/or existing site contamination can be exposed to poor quality outside air that is used for ventilation. Refer to the following credits:

- SSc4 Alternative Transportation
- SSc3 Brownfield Redevelopment

IAQ can be positively affected if teams specify materials and furnishings that do not release harmful or irritating chemicals, including Volatile Organic Compounds (VOCs) from paints and solvents. Refer to the following credits:

- IEQp2 Environmental Tobacco Smoke (ETS) Control
- IEQc4 Low-Emitting Materials
- IEQc5 Indoor Chemical and Pollutant Source Control

IEQp2 ENVIRONMENTAL TOBACCO SMOKE (ETS) CONTROL

Intent

This prerequisite seeks to prevent or minimize the exposure of environmental tobacco smoke to building occupants, indoor surfaces, and ventilation air distribution systems. Prohibiting smoking of cigarettes, pipes, and cigars is the most effective way to avoid ETS-related health problems. If this is not possible, designated smoking areas with separate ventilation systems to isolate the space from the rest of the building can be provided.

Requirements

Teams have two options to achieve this credit:

Option 1: Prohibit smoking in the building.

Designate smoking areas outside of the building located at least 25 feet from building entries, outdoor air intakes, and operable windows. Provide signage directing building occupants to areas where smoking is allowed.

Option 2: Prohibit smoking in the building except in designated smoking rooms.

Designate smoking areas outside of the building located at least 25 feet from building entries, outdoor air intakes, and operable windows. Provide signage directing building occupants to areas where smoking is allowed.

Designated smoking areas within buildings must comply with the following requirements:

- Directly exhausts to outside the building
- No recirculation of air supplied to the room
- Room to be at negative pressure
- Impermeable deck-to-deck partitions

Referenced Standards

The following are two referenced standards for this prerequisite:

1. ANSI/ASTM—E779—03 Standard Test Method for Determining Air Leakage Rate By Fan Pressurization. Download this standard at *www.astm.org*.
2. Residential Manual for Compliance with California's 2001 Energy Efficiency Standards (for low-rise residential buildings, Chapter 4). Download this standard at *www.energy.ca.gov/title24/2005standards/residential_manual.html*.

Related Credits

If the team is providing smoking areas within the building, the spaces will need separate ventilation systems. These spaces will require additional energy, commissioning, and measurement and verification, as addressed in the following prerequisites and credits:

- EAp1 Fundamental Commissioning of Building Energy Systems
- EAc1 Optimize Energy Performance
- EAc3 Enhanced Commissioning
- EAc5 Measurement and Verification

IAQ performance is affected by both indoor and outdoor smoking. The following prerequisites and credits will need to take the potential ETS contamination into consideration:

- IEQp1 Minimum Indoor Air Quality Performance
- IEQc1 Outdoor Air Delivery Monitoring
- IEQc2 Increased Ventilation

IEQc1 OUTDOOR DELIVERY MONITORING

Intent

The focus of this credit is to sustain occupant comfort and well-being by providing capacity for monitoring of ventilation systems. Measuring CO_2 concentration helps to determine air-change effectiveness. An elevated CO_2 level indicates poor air quality and a potential build-up of indoor air pollutants. CO_2 by itself is not harmful; however, in high concentrations, it displaces oxygen and can lead to headaches, dizziness, and increased heart rate.

Heating, ventilating, and air conditioning (HVAC) systems are designed to flush out indoor airborne contaminants by replacing old indoor air with outdoor air. Many conventional ventilation systems do not directly measure the amount of outside air that is delivered. Installing permanent monitoring systems will enable the ventilation systems to perform as designed.

Airflow and CO_2 monitoring systems can be applied to any building or HVAC system whether it is mechanically or naturally vented. This allows building personnel to be alerted when the system is not operating as designed; in addition, automated control systems can inform them about necessary operation adjustments.

Requirements

To achieve this credit, a permanent ventilation monitoring system must be installed in the building to verify that the ventilation system is operating as designed. The monitoring system must sound an alarm if airflow values or the level of CO_2 vary by 10 percent or more from the design values, and must comply with one of the following:

<u>Case 1</u>: Mechanically Ventilated Spaces

In all densely occupied spaces where the density is 25 people or more per 1,000 square feet, monitor CO_2 concentrations. Position the CO_2 monitors between 3' and 6' above the floor. Also, provide a direct outdoor airflow measurement device that can measure the minimum outdoor air intake flow with an accuracy of ±15 percent of the design minimum outdoor intake, as per ASHRAE 62.1—2007 with errata.

<u>Case 2</u>: Naturally Ventilated Spaces

In all naturally ventilated spaces, monitor CO_2 concentrations. CO_2 monitors must be between 3' and 6' above the floor.

Referenced Standard

The referenced standard for this credit is ASHRAE 62.1—2007: Ventilation for Acceptable Indoor Air Quality. Download this standard at *www.ashrae.org*.

Related Credits

Monitoring air flow can enhance energy performance, commissioning, and measurement, and verification efforts for the following credits:

- EAp1 Fundamental Commissioning of Building Energy Systems
- EAc3 Enhanced Commissioning
- EAc5 Measurement and Verification
- IEQc2 Increased Ventilation

IEQc2 INCREASED VENTILATION

Intent

This credit focuses on increasing ventilation by providing increasing volumes of outside air to improve indoor air quality and enhance occupant comfort, well-being, and productivity.

"Americans spend about 90 percent of their time indoors, where concentrations of pollutants are often much higher than those outside. Risk assessments performed for radon, environmental tobacco smoke (ETS), and lead have shown that health risks are substantial. Thousands of chemicals and biological pollutants are found indoors, many of which are known to have significant health impacts, both indoors and in other environments."[1]

Increasing ventilation levels improves indoor air quality, which benefits occupant health and productivity.

Requirements

To achieve this credit, teams must complete one of the following compliance paths, plus fulfill one of the two options to demonstrate compliance with this credit.

Path 1: Mechanically Ventilated Spaces

Design mechanical ventilation to exceed the minimum rates required by ASHRAE Standard 62.1-2—2007 for breathing zone outdoor air ventilation rates in all occupied spaces by at least 30 percent.

Path 2: Naturally Ventilated Spaces

For spaces that have been designed to meet the Carbon Trust "Good Practice Guide 237" 1998, design of the naturally ventilated spaces must follow the flow diagram process shown in Figure 1.18 of the Chartered Institution of Building Services Engineers (CIBSE), Applications Manual 10, 2005: *Natural Ventilation in Non-Domestic Buildings*.

In addition, teams must comply with one of the following options:

Option 1: Using diagrams and calculations, show that the natural ventilation systems design meets the recommendations set forth in the CIBSE Applications Manual 10, 2005: *Natural Ventilation in Non-Domestic Buildings*.

Option 2: Use a macroscopic, multizone, analytic model to predict that the room-by-room airflows will effectively ventilate naturally for at least 90 percent of occupied spaces, defined as providing the minimum ventilation rates required by ASHRAE 62.1—2007, Chapter 6, with errata, but without addenda.

Referenced Standard

The referenced standard for this credit is ASHRAE 62.1—2007: Ventilation for Acceptable Indoor Air Quality. Download this standard at *www.ashrae.org*.

Related Credits

Providing increased ventilation can affect energy performance, and require additional commissioning and measurement and verification efforts for the following related credits:

- EAp1 Fundamental Commissioning of Building Energy Systems
- EAp2 Minimum Energy Performance
- EAc1 Optimize Energy Performance
- EAc3 Enhanced Commissioning
- EAc5 Measurement and Verification
- IEQc1 Outdoor Air Delivery Monitoring

IEQc3.1 CONSTRUCTION IAQ MANAGEMENT PLAN—DURING CONSTRUCTION

Intent

The focus of this credit is to sustain contractor and building occupant health by reducing IAQ problems resulting from construction activities. During demolition and construction, a wide variety of pollutants are released in the air, which can result in poor IAQ that continues through the lifetime of the building. The implementation of a construction IAQ management plan can minimize construction-related IAQ problems.

Requirements

To achieve this credit, teams must develop and implement a construction IAQ management plan for the construction and preoccupancy phases of the project according to the following requirements:

- During construction, meet or exceed the control measures recommended by Sheet Metal and Air Conditioning National Contractors Association (SMACNA) *IAQ Guidelines for Occupied Buildings Under Construction*, 2nd Edition 2007, ANSI/SMACNA 008-2008 (Chapter 3).
- Protect stored on-site and installed absorptive materials from moisture damage.
- Use filtration media with a minimum efficiency reporting value (MERV) of 8 at each return air grille for all permanently installed air handlers operated during construction. Replace filtration media immediately prior to occupancy.

Referenced Standard

The referenced standard for this credit is the Sheet Metal and Air Conditioning National Contractors Association (SMACNA) *IAQ Guidelines for Occupied Buildings Under Construction*, 2nd Edition 2007, ANSI/SMACNA 008-2008 (Chapter 3). Download this standard at *www.smacna.org*.

Related Credits

The building IAQ can be affected long after construction is completed. Implementing the construction IAQ management plan, selecting low VOC finish materials and isolating indoor sources of pollution can reduce future IAQ issues and are addressed by the following credits:

- IEQc3.2 Construction IAQ Management Plan—Before Occupancy
- IEQc4 Low-Emitting Materials
- IEQc5 Indoor Chemical and Pollutant Source Control

IEQc3.2 CONSTRUCTION IAQ MANAGEMENT PLAN—BEFORE OCCUPANCY

Intent

This credit focuses on the period of time after all construction work, including punch-list items and finishes, and cleaning operations are completed, and before the building occupants move in. To ensure the building is free of major contaminants, and concentrations are below recognized acceptable levels before occupancy, teams can use either the building flush-out procedure, or IAQ testing. The flush-out procedure uses the building HVAC system to evacuate airborne contaminants. The IAQ testing procedure uses the IAQ baseline to confirm that major contaminants are below acceptable levels.

Requirements

To achieve this credit, teams can choose to perform a building flush-out, or IAQ testing as described in the following two options:

Option 1: Building Flush-Out

Path 1: Pre-occupancy Flush-Out

Perform flush-out after all construction, including punch-list and finishes, and all cleaning operations are completed. All filtration is to be replaced. Building is to be flushed with outside air supplied at a total air volume of 14,000 cubic feet of outdoor air per square foot of floor area while maintaining an internal building temperature of at least 60° F, and a relative humidity no higher than 60 percent.

Commissioning can occur during the flush-out procedure if no contaminants are introduced into the building.

Path 2: Post-occupancy Flush-Out

Prior to occupancy, the space must be flushed out with a minimum 3,500 cubic feet of outdoor air per square foot of floor area. Then, after the space is occupied, ventilate the building at a minimum rate of 0.30 cubic feet per minute (CFM) per square foot of outside air, or at the design minimum outside air rate determined in IEQp1 Minimum Indoor Air Quality Performance, whichever is greater. Ventilation must begin at least three hours prior to occupancy and continue during occupancy for each day of the flush-out period. This procedure must be maintained until a total of 14,000 cubic feet of outdoor air per square foot of floor area has been delivered.

Commissioning can occur during the flush-out procedure if no contaminants are introduced into the building.

Option 2: Air Testing

Conduct baseline IAQ testing after all construction, including punch-list and finishes, and all cleaning operations are completed, but prior to occupancy. The baseline IAQ testing is to be conducted according to the EPA Compendium of Methods for the Determination of Air Pollutants in Indoor Air and USGBC Requirements addressed in the 2009 *LEED® Reference Guide for Green Building Design and Construction.*

Air sampling must be conducted prior to occupancy during normal business hours according to the following criteria:

- Building ventilation system to be started at the normal daily start time and operated at the minimum outside air flow rate for the duration of the test.
- All interior finishes must be installed including, but not limited to, carpet and acoustical ceiling tile.

- Air samples must be collected in the breathing zone between 3' and 6' above the floor over at least a four-hour period.
- Provide sampling locations for each portion of the building served by a separate ventilation system, and at least one sampling point per 25,000 square feet or for each contiguous floor area, whichever is larger.
- For each sampling point where the maximum concentration for a contaminant is exceeded, conduct an additional flush-out with outside air, and retest. Repeat until all concentrations are below established limits.

Movable furnishing, such as workstations and partitions, is not required to be installed at the time of building flush-out or air testing; however, it is recommended that those furnishings be installed.

Referenced Standard

The referenced standard for this credit is the U.S. EPA Compendium of Methods for Determination of Air Pollutants in Indoor Air.

Related Credits

The construction management plan required by this credit is also required by IEQc3.1 Construction IAQ Management Plan—During Construction.

Filtration media, interior building components and finishes can affect indoor air quality and test results. Refer to the following related credits:

- IEQc4 Low-Emitting Materials
- IEQc5 Indoor Chemical and Pollutant Source Control

Typically, indoor air contaminants can be diluted by introducing outside air. Ventilation rates are addressed in the following prerequisite and credit:

- IEQp1 Minimum Indoor Air Quality Performance
- IEQc2 Increased Ventilation

IEQc4 LOW-EMITTING MATERIALS—INTRODUCTION

Intent

Low-emitting material refers to products and installation processes that have the potential to adversely affect the IAQ of a space and its occupants by exposing them to the off-gassing of contaminants from those materials.

Indoor contaminant sources include all surfaces in contact with indoor air. Visible surfaces, such as flooring, ceilings, interior furnishings, and walls, are the easiest potential contaminant sources to identify. Ventilation systems, their components, wall and ceiling cavity components, as well as caulking for windows and wall insulation, can also be indoor air contaminant sources.

The following series of summarized credits seeks to protect the comfort and well-being of installers and occupants by reducing the quantity of indoor air contaminants that contain odors, irritants and/or are harmful.

IEQc4.1 Low-Emitting Materials—Adhesives and Sealants

IEQc4.2 Low-Emitting Materials—Paints and Coatings

IEQc4.3 Low-Emitting Materials—Flooring Systems

IEQc4.4 Low-Emitting Materials—Composite Wood and Agrifiber Products

IEQc4.1 LOW-EMITTING MATERIALS—ADHESIVES AND SEALANTS

Intent

Volatile Organic Compounds (VOCs) are some of the most recognized compounds negatively affecting indoor air quality and the Earth's atmosphere. VOCs also contribute to ground-level ozone, which is a major component of smog.

"Volatile organic compounds (VOCs) are emitted as gases from certain solids or liquids. VOCs include a variety of chemicals, some of which may have short- and long-term adverse health effects. Concentrations of many VOCs are consistently higher indoors (up to ten times higher) than outdoors. VOCs are emitted by a wide array of products numbering in the thousands. Examples include: paints and lacquers, paint strippers, cleaning supplies, pesticides, building materials and furnishings, office equipment, such as copiers and printers, correction fluids and carbonless copy paper, graphics and craft materials, including glues and adhesives, permanent markers, and photographic solutions."[2]

Requirements

To achieve this credit, all the VOC content of adhesives and sealants must comply with the South Coast Air Quality Management District (SCAQMD) Rule #1168.

Referenced Standard

There are two referenced standards for this credit as listed below.

1. Green Seal Standard 36 (GS-36), for commercial adhesives, effective October 19, 2000. Download this standard at *www.greenseal.org/certification/standards/commercial_adhesives_GS_36.cfm.*
2. South Coast Air Quality Management District (SCAQMD) Rule #1168. Download this standard at *www.aqmd.gov/rules/reg/reg11/r1168.pdf.*

Related Credits

The following prerequisite and credit also address air quality affected by sources from within the building:

- IEQc4.2 Low-Emitting Materials—Paints and Coatings
- IEQc4.3 Low-Emitting Materials—Flooring Systems
- IEQc4.4 Low-Emitting Materials—Composite Wood and Agrifiber Products
- IEQp2 Environmental Tobacco Smoke (ETS) Control
- IEQc5 Indoor Chemical and Pollutant Source Control

IEQc4.2 LOW-EMITTING MATERIALS—PAINTS AND COATINGS

Intent

"Organic chemicals are widely used as ingredients in household products. Paints, varnishes, and wax all contain organic solvents, as do many cleaning, disinfecting, cosmetic, degreasing, and hobby products. Fuels are made up of organic chemicals. All of these products can release organic compounds while you are using them, and, to some degree, when they are stored."[2]

Requirements

To achieve this credit, all paint and coatings applied to the interior of the building must comply with the following requirements:

- VOC emissions for paints, primers, and coatings applied to interior walls must not exceed the VOC content limits of Green Seal Standard GS-11, paints, 1993 first edition.
- VOC content for anti-corrosive and anti-rust paints applied to interior metal substrates are not to exceed 250g/L, as established by Green Seal Standard GC-03, anti-corrosive paints, 1997 second edition.
- VOC control limits for clear wood shellacs applied to interior elements must comply with SCAQMD Rule 1113, architectural coatings, 2004.

Referenced Standards

There are three referenced standards for this credit as listed below.

1. Green Seal Standards GC-03, anti-corrosive and anti-rust paints. Download this standard at *www.greenseal.org/certification/standards/anti-corrosivepaint.pdf.*
2. Green Seal Standards GS-11, commercial flat and non-flat paints. Download this standard at *www.greenseal.org/certification/standards/paints_and_coatings.pdf.*
3. South Coast Air Quality Management District (SCAQMD) Rule #1113. Download this standard at *www.aqmd.gov/rules/reg/reg11/r1113.pdf.*

Related Credits

The following prerequisite and credits also address air quality affected by sources from within the building:

- IEQc4.1 Low-Emitting Materials—Adhesives and Sealants
- IEQc4.3 Low-Emitting Materials—Flooring Systems
- IEQc4.4 Low-Emitting Materials—Composite Wood and Agrifiber Products
- IEQp2 Environmental Tobacco Smoke (ETS) Control
- IEQc5 Indoor Chemical and Pollutant Source Control

IEQc4.3 LOW-EMITTING MATERIALS—FLOORING SYSTEMS

Intent

"The vast majority of current conventional flooring products have little to do with eco-friendliness or even aesthetic appeal. Approximately 70 percent of floor covering in the United States is carpet, mostly made from nylon, which is a familiar term that masks its ultimate source: petroleum oil. Besides being made from one of our most endangered resources, synthetic carpet is high in volatile organic compound (VOC) off-gassing, provides a safe haven for many allergens, and, as long-term homeowners know, wears with very little grace. Once removed and disposed, nylon carpet is fated for a non-biodegradable repose in a landfill. Other popular floors have similar issues: the petroleum-based vinyl flooring that is common in kitchen and bathrooms everywhere is made of polyvinyl chlorate (PVC), which contains several chemicals called phthalates. Phthalates have raised enough health concerns, particularly with respect to male development, that in 2005, the European Union banned use of some phthalates in children's toys."[3]

Requirements

To achieve this credit, teams can choose one of the following compliance paths:

Option 1:

- All carpets must meet Carpet and Rug Institute's Green Label Plus program.
- All carpet cushions must meet Carpet and Rug Institute's Green Label program.
- All carpet adhesive must have a VOC limit of 50g/L as per EQc4.1.
- All vinyl, linoleum, laminate, wood, ceramic, and rubber flooring, as well as wall base, must meet FloorScore's standard by independent third-party certification.
- All concrete, wood, bamboo, and cork floor finishes, such as sealer, stain, and finish, must meet the requirements of SCAQMD rule #1113.
- All tile setting adhesives and grout must meet SCAQMD rule #1168.

Option 2: All interior flooring elements must meet the testing and product requirements of the California Department of Health Services Standard Practice for the Testing of VOC Emissions from Various Sources Using Small Scale Environmental Chambers, including 2004 addenda.

Referenced Standards

There are six referenced standards for this credit as listed below.

1. Carpet and Rug Institute (CRI) Green Label Plus Testing Program. Download this standard at *www.carpet-rug.com*.
2. South Coast Air Quality Management District (SCAQMD) Rule #1113. Download this standard at *www.aqmd.gov/rules/reg/reg11/r1113.pdf*.
3. South Coast Air Quality Management District (SCAQMD) Rule #1168. Download this standard at *www.aqmd.gov/rules/reg/reg11/r1168.pdf*.
4. FloorScore™ Program Resilient Floor Covering Institute. Download this standard at *www.rfci.com/int_floorScore.htm*.
5. California Department of Health Services Standard Practice for the Testing of VOC Emissions from Various Sources Using Small Scale Environmental Chambers, including 2004 addenda. California Department of Health Services. Download this standard at *www.cal-iaq.org/VOC/Section01350_15_2004_FINAL_PLUS_ADDENDUM-2004-01.pdf*.
6. State of California Standard 1350, Section 9, Standard Practice for the Testing of VOC Emissions from Various Sources Using Small Scale Environmental Chambers, testing criteria. Download this standard at *www.dhs.ca.gov.ps.deodc/ehlb/iaq/VOCS/Section1350_7_15_2004_FINAL_PLUS_ADDENDUM-2004-01.pdf*.

Related Credits

The following prerequisite and credit also address air quality affected by sources from within the building:

- IEQc4.1 Low-Emitting Materials—Adhesives and Sealants
- IEQc4.2 Low-Emitting Materials—Paints and Coatings
- IEQc4.4 Low-Emitting Materials—Composite Wood and Agrifiber Products
- IEQp2 Environmental Tobacco Smoke (ETS) Control
- IEQc5 Indoor Chemical and Pollutant Source Control

IEQc4.4 LOW-EMITTING MATERIALS—COMPOSITE WOOD AND AGRIFIBER PRODUCTS

Intent

Composite wood and agrifiber products are popular because they can be used in a variety of ways, often in applications similar to solid wood products. In some applications, these alternative materials are preferred because they can be designed to meet application-specific performance requirements. Composite wood is made up of fibers from small diameter trees, small pieces of wood, and wood that has defects. Composite wood products are often stronger and less prone to humidity-induced warping than equivalent solid woods, although most particle and fiber-based boards readily soak up water unless they are treated with sealant or paint. Agrifiber is comprised of materials including crop wastes, such as grass straw, cereal straw, rice straw, bagasse, stover and stalks, perennial native grasses, weeds and ditch grass, wood wastes, paper and cardboard that is blended, chopped, dried, and milled.

Composite wood and agrifiber products include particleboard, medium density fiberboard (MDF), plywood, wheatboard, strawboard, panel substrates, and door cores.

Requirements

To achieve this credit, composite wood and agrifiber products used in the interior of the building (for example, inside the weatherproofing system) cannot contain added urea-formaldehyde resins. The laminating adhesives used for on-site and shop-applied fabrications of composite wood and agrifiber assemblies cannot contain added urea-formaldehyde resins.

Items such as fixtures, furniture, and equipment (FF&E) are not included in the calculations for this credit, because they are not considered part of the base building.

Related Credits

The following prerequisite and credit also address air quality affected by sources from within the building:

- IEQc4.1 Low-Emitting Materials—Adhesives and Sealants
- IEQc4.2 Low-Emitting Materials—Paints and Coatings
- IEQc4.3 Low-Emitting Materials—Flooring Systems
- IEQp2 Environmental Tobacco Smoke (ETS) Control
- IEQc5 Indoor Chemical and Pollutant Source Control

IEQc5 INDOOR CHEMICAL AND POLLUTANT SOURCE CONTROL

Intent

The focus of this credit is to limit the amount of particulate, chemical, and biological contaminants to which building occupants are exposed by reducing or mitigating the sources of the contaminants. This can be easily accomplished by installing entryway systems at building entry points and isolating areas where chemicals are used.

Requirements

To attain this credit, teams must minimize chemical pollutant cross-contamination of regularly occupied areas using the following methods:

- Install permanent entryway systems (grills, grates, etc.) to capture dirt, particulate, etc. at all high-volume building entryways.
- Where chemical use occurs, including housekeeping areas and copying/printing rooms, provide self-closing doors and deck-to-deck partitions with separate outside exhaust at a rate of at least 0.50 cubic feet per minute per square foot. Ensure no air recirculation, and maintain the room at negative pressure.
- If the building is mechanically ventilated, replace outside air supply and return air filtration media with a Minimum Efficiency Reporting Value (MERV) rating of at least 13 prior to occupancy.
- In spaces such as housekeeping, janitorial, and science laboratories, provide appropriate disposal of hazardous liquid waste where mixture of water and chemical concentrate occurs.

Referenced Standard

The referenced standard for this credit is ANSI/ASHRAE52.2—1999: Method of testing general ventilation air-cleaning devices for removal efficiency by particle size. Download this standard at *www.ashrae.org*.

Related Credits

This credit requires high-efficiency filtration media. Coordinate filtration requirements with the Construction IAQ Management Plan addressed in the following credits:

- IEQc3.1 Construction IAQ Management Plan—During Construction
- IEQc3.2 Construction IAQ Management Plan—Before Occupancy

The exhaust requirements for this credit can require additional fan energy. Building energy systems and commissioning are addressed in the following prerequisites and credits:

- EAp1 Fundamental Commissioning of Building Energy Systems
- EAp2 Minimum Energy Performance
- EAc1 Optimize Energy Performance
- EAc3 Enhanced Commissioning

The ventilation and filtration media requirements of this credit should be coordinated with the following prerequisite and credit:

- IEQp1 Minimum Indoor Air Quality Performance
- IEQc1 Outdoor Air Delivery Monitoring

IEQc6.1 CONTROLLABILITY OF SYSTEMS—LIGHTING

Intent

This credit focuses on building lighting systems and providing building occupants the ability to control lighting levels in their workspace to increase productivity, comfort, and well-being.

Lighting design should be integrated into space planning, lighting schemes, lighting controls, task lighting, and HVAC early in the design process. Evaluation of window design to provide optimal size, orientation, and aspect ratio for vision and/or daylight functions desired for windows can also be beneficial to the lighting of the space.

Requirements

To achieve this credit, teams must provide individual lighting controls for 90 percent of building occupants, and provide lighting system controls for multi-occupant spaces that are able to adjust to the needs of the group.

Related Credits

Lighting systems are affected by many design components of the building addressed by the following credits:

- EAp1 Fundamental Commissioning of Building Energy Systems
- EAp2 Minimum Energy Performance
- EAc1 Optimize Energy Performance
- EAc3 Enhanced Commissioning
- IEQc6.2 Controllability of Systems—Thermal Comfort
- EAc8 Daylight and Views

IEQc6.2 CONTROLLABILITY OF SYSTEMS—THERMAL COMFORT

Intent

This credit focuses on building thermal systems and providing building occupants the ability to control thermal levels in their workspace to increase productivity, comfort, and well-being.

Requirements

To achieve this credit, teams must provide individual comfort controls for 50 percent of building occupants. Operable windows are an acceptable alternative for comfort controls if building occupants are located 20 feet inside, and 10 feet to either side of an operable window. The areas of operable window must meet the requirement of ASHRAE 62.1—2007, paragraph 5.1 Natural Ventilation (with errata, but without addenda).

In addition, comfort system controls must be provided for all shared multi-occupant spaces to enable adjustments that meet group needs and preferences.

Referenced Standard

There are two referenced standards for this credit as listed below:

1. ASHRAE 62.1—2007: Ventilation for Acceptable Indoor Air Quality. Download this standard at *www.ashrae.org*.
2. ASHRAE 55—2004: Thermal Environmental Conditions for Human Occupancy. Download this standard at *www.ashrae.org*.

Related Credits

Thermal comfort systems and their controllability are affected by many design components of the building that are related to the following credits:

- EAp1 Fundamental Commissioning of Building Energy Systems
- EAp2 Minimum Energy Performance
- EAc1 Optimize Energy Performance
- EAc3 Enhanced Commissioning
- EAc5 Measurement and Verification
- IEQc5 Indoor Chemical and Pollutant Source Control
- IEQc6.1 Controllability of Systems—Lighting
- IEQc8 Daylight and Views

IEQc7.1 THERMAL COMFORT—DESIGN

Intent

This credit supports the productivity and well-being of building occupants by providing a thermally comfortable environment.

Teams should establish temperature and humidity comfort ranges and design the building envelope and HVAC system to maintain these ranges. Human thermal comfort can be controlled through air temperature, air velocity, humidity levels, controlling air infiltration of the building, and using shading devices, insulation, and thermal mass to manage interior surface temperatures.

Requirements

To achieve this credit, teams need to design the building HVAC systems and building envelope to comply with ASHRAE Standard 55—2004, section 6.1.1.

Referenced Standard

The referenced standard for this credit is ASHRAE 55—2004: Thermal Environmental Conditions for Human Occupancy. Download this standard at *www.ashrae.org*.

Related Credits

The following prerequisite and credits address the thermal comfort of building occupants, including air temperature, radiant temperature, relative humidity, and air speed:

- EAp2 Minimum Energy Performance
- EAc1 Optimize Energy Performance
- EAc5 Measurement and Verification

The following prerequisite and credit address commissioning of thermal comfort systems:

- EAp1 Fundamental Commissioning of Building Energy Systems
- EAc3 Enhanced Commissioning

The following prerequisite and credits address occupant comfort:

- IEQp1 Minimum Indoor Air Quality Performance
- IEQc2 Increased Ventilation
- IEQc6.2 Controllability of Systems—Thermal Comfort
- IEQc7.2 Thermal Comfort—Verification

IEQc7.2 THERMAL COMFORT—VERIFICATION

Intent

Receiving feedback from building occupants can help the building personnel adjust thermal comfort to minimize occupant complaints. The survey can be performed by either facility operators or outside consultants and should focus on satisfaction with the thermal environment. There should also be a plan to make corrective actions to problems identified in the survey.

Requirements

To achieve this credit, teams must agree to conduct a thermal comfort survey of building occupants within 6 to 18 months following occupancy. The survey should collect anonymous responses regarding thermal comfort in the building. A corrective action plan should be developed to address problems identified by more than 20 percent of occupants.

Referenced Standard

The referenced standard for this credit is ASHRAE 55—2004: Thermal Comfort Conditions for Human Occupancy. Download this standard at *www.ashrae.org*.

Related Credits

The following prerequisite and credits address the thermal comfort of building occupants, including air temperature, radiant temperature, relative humidity, and air speed:

- EAp1 Fundamental Commissioning of Building Energy Systems
- EAc3 Enhanced Commissioning
- EAc5 Measurement and Verification

The following prerequisite and credits address occupant comfort:

- IEQp1 Minimum Indoor Air Quality Performance
- IEQc2 Increased Ventilation
- IEQc6.2 Controllability of Systems—Thermal Comfort
- IEQc7.1 Thermal Comfort—Design

IEQc8.1 DAYLIGHT AND VIEWS—DAYLIGHT

Intent

This credit focuses on providing daylighting to regularly occupied areas of the building to connect indoor and outdoor spaces. Daylighting can reduce the need for interior electric lighting and reduce energy use.

Teams pursuing this credit should carefully review the balance between heat gain and loss, glare control, visual quality, and variation in daylight availability. Teams can use shading devices, light shelves, courtyards, atriums, and window glazing to achieve their daylighting goals. Other considerations should include the building orientation, window sizes and spacing, glazing selection, interior finishes, and locations of interior walls.

Requirements

To attain this credit, teams must demonstrate credit compliance through one of the following four options:

Option 1: Daylight Simulation Model

Through computer simulation, demonstrate that a minimum daylight illumination level of 25 foot-candles (fc) and a maximum level of 500 fc in clear sky conditions is achieved in 75 percent of regularly occupied areas. Illumination levels to be calculated on September 21st at 9:00 AM and 3:00 PM.

Option 2: Prescriptive

Daylight Zone—Side-Lighting

Use the following calculation to determine if the visible light transmittance (VLT) and window-to-floor area ratio (WFR) of the daylight zone is between 0.150 and 0.180.

$$0.150 < VLT \times WFR < 0.180$$

For this calculation, the ceiling cannot obstruct the space when a line is drawn diagonally from the head of the window and extends into the space a distance that is twice the height of the window, as illustrated in the following graphic.

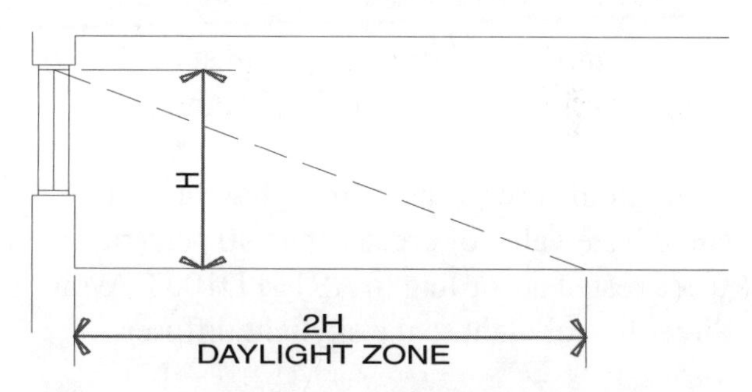

Figure 6.1 Daylight Zone

To ensure daylight effectiveness, provide sunlight redirection and/or glare control devices.

Daylight Zone—Top-Lighting

The daylit zone is the zone directly under the skylight, plus, in each direction, the lesser of the three following conditions:

- Seventy percent of the ceiling height

OR

- One-half the distance to the edge of the nearest skylight

OR

- The distance to any permanent opaque partition that is farther than 70 percent of the distance between the top of the partition and the ceiling

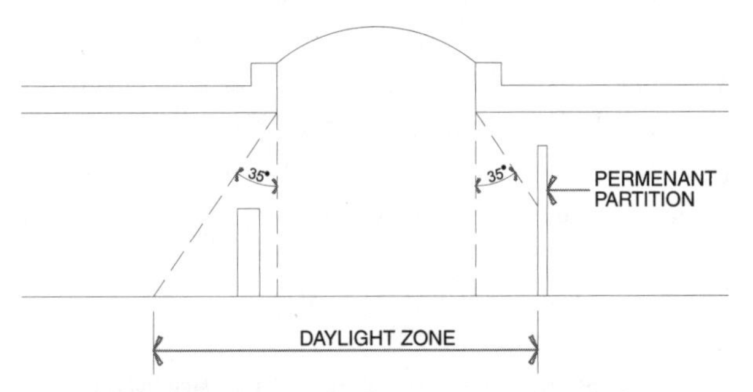

Figure 6.2 Daylight Zone-Top Lighting

Achieve skylight roof coverage between 3 percent and 6 percent of the roof area with a minimum 0.5 VLT.

The distance between skylights must not be more than 1.4 times the ceiling height.

Skylight diffuser, if used, must have a measured haze value of greater than 90 percent when tested according to ASTM D1003. Avoid direct line of sight to the skylight diffuser.

Exceptions for areas where tasks would be hindered by the use of daylight will be considered on their merits.

Option 3: Measurement

Through records of daylight measurement, demonstrate that a minimum daylight illumination level of 25 foot-candles is achieved in 75 percent of regularly occupied areas. Measurements must be taken on a 10-foot grid for all occupied spaces and recorded on floor plans.

Only square footage associated with the portion of rooms or spaces meeting the minimum illumination requirements may be counted in the calculations.

In all cases, daylight redirection and/or control devices are to be provided to avoid high-contrast situations that could impede visual tasks.

Exceptions for areas where tasks would be hindered by the use of daylight will be considered on their merits.

Option 4: Combination

Any of the previously mentioned calculation methods may be combined to document the minimum daylight illumination in at least 75 percent of all regularly occupied space. The different methods used in each space must be clearly recorded on all building plans.

In all cases, only the square footage associated with the portions of rooms or spaces meeting the requirements may be applied toward the total area calculation required to qualify for this credit.

In all cases, provide glare control devices to avoid high-contrast situations that could impede visual tasks.

Exceptions for areas where tasks would be hindered by the use of daylight will be considered on their merits.

Referenced Standard

The referenced standard for this credit is ASTM D1003-07e1, Standard Test Method for Haze and Luminous Transmittance of Transparent Plastics.

Related Credits

Increasing the area of vision glazing can provide more access to views and may contribute to the following credit:

- IEQc8.2 Daylight and Views—Views

Increasing the window-to-wall ratio can affect the energy performance, and lighting design strategies to conserve energy through the use of daylighting controls. Refer to the following prerequisite and credits:

- EAp2 Minimum Energy Performance
- EAc1 Optimize Energy Performance
- IEQc6 Controllability of Systems

Exemplary Performance Credit

One innovation credit point is available in Innovation in Design for teams demonstrating 95 percent daylighting.

IEQc8.2 DAYLIGHT AND VIEWS—VIEWS

Intent

This credit focuses on providing a visual connection between indoor and outdoor spaces.

Requirements

To achieve this credit, teams need to demonstrate that a direct line of sight to vision glazing between 2'–6" and 7'–6" above finished floor (AFF) for building occupants in 90 percent of all regularly occupied spaces.

To determine the area with a direct line of sight, total the square footage of the regularly occupied spaces that meet the following criteria:

- In plan view, the area is within sight lines drawing from perimeter vision glazing.
- In section view, a direct sight line at 24" AFF can be drawn from the area to perimeter vision glazing.

The line of sight may be drawn through interior glazing. For private offices, the entire square footage of the office may be counted if 75 percent or more of the area has a direct line of sight to perimeter vision glazing. For classrooms and other multi-occupant spaces, the actual square footage with a direct line of sight to perimeter vision glazing is counted.

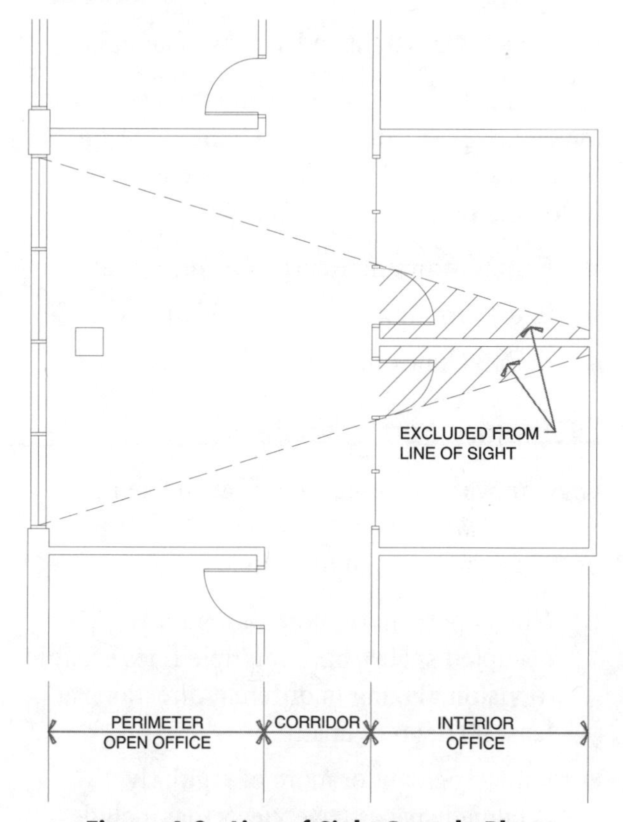

Figure 6.3 Line of Sight Sample Plan

Regularly occupied spaces include office spaces, conference rooms, classrooms, core learning spaces, and cafeterias. Areas that do not need to be considered include support areas for copying, storage, mechanical equipment, laundry, and restrooms.

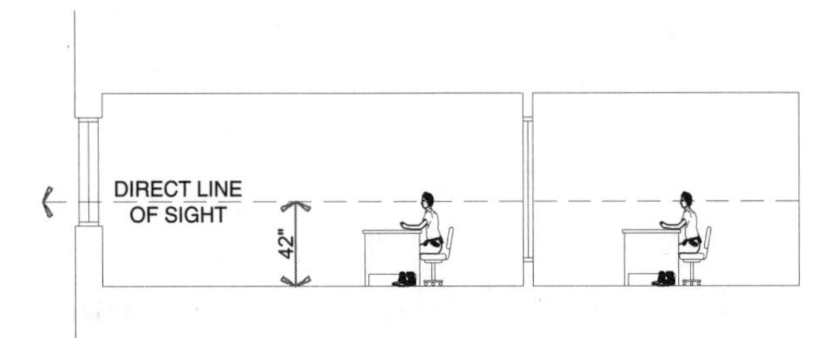

Figure 6.4 Direct Line of Sight Through Interior Window Over Low Partition

Related Credits

Increasing the area of vision glazing can provide more access to daylight and may contribute to the following credit:

- IEQc8.1Daylight and Views—Daylight

Increasing the window-to-wall ratio can affect the energy performance, and lighting design strategies may conserve energy. Refer to the following prerequisite and credits:

- EAp2 Minimum Energy Performance
- EAc1 Optimize Energy Performance
- IEQc6 Controllability of Systems

Exemplary Performance

One innovation credit point is available in Innovation in Design for projects meeting two of the following requirements:

- Ninety percent or more of regularly occupied spaces have multiple lines of sight to vision glazing in different directions at least 90 degrees apart.
- Ninety percent or more of regularly occupied spaces have views that include views of at least two of the following three options:
    1. Vegetation
    2. Human activity
    3. Objects at least 70 feet from the exterior of the glazing
- Ninety percent or more of regularly occupied spaces have access to unobstructed viewslocated within a distance of three times the head height of the vision glazing.
- Ninety percent or more of regularly occupied spaces have access to views with a view factor of three or greater.

ABBREVIATIONS AND ACRONYMS

4-PCH	4-phenylcylohene
AFF	Above Finished Floor
CO_2	Carbon Dioxide
E_{ac}	Air Change Effectiveness
ETS	Environmental Tobacco Smoke
FF&E	Fixtures, Furniture, and Equipment
SCAQMD	South Coast Air Quality Management District
HVAC	Heating Ventilating and Air Conditioning System
IAQ	Indoor Air Quality
MERV	Minimum Efficiency Reporting Value
T_{vis}	Visible Transmittance
VLT	Visible Light Transmittance
VOCs	Volatile Organic Compounds
WFT	Window-to-Floor Ratio

FOOTNOTES

1. U.S. Environmental Protection Agency, *Why Study Human Health Indoors?*, 2006, http://www.epa.gov/iaq/hbhp/hbhp_report.pdf.
2. U.S. Environmental Protection Agency, *An Introduction to Indoor Air Quality, Organic Gases (Volatile Organic Compounds—VOCs)*, 2009, http://www.epa.gov/iaq/voc.html.
3. Green Living Ideas, Inc., *Green Flooring for Sustainable Spaces,* 2009, http://greenlivingideas.com/flooring/green-flooring-for-sustainable-spaces.

LESSON 6 QUIZ

1. What is the referenced standard for IEQp1 Minimum Indoor Air Quality Performance?

   A. ASHRAE 55—2004

   B. ASHRAE 62.1—2007

   C. ASHRAE 129—1999

   D. ASHRAE 52.5—1999

   E. ASHRAE 90.1—2007

2. At what height does IEQc1 Outdoor Delivery Monitoring require CO_2 monitors to be placed throughout the building?

   A. Between 33" and 66" above the floor

   B. Between 3' and 7' above the floor

   C. Between 3' and 6' above the floor

   D. Between 30" and 60" above the floor

   E. Between 2'–6" and 7'–6" above the floor

3. What are the two options to confirm that major contaminants are below acceptable levels to earn IEQc3.2 Construction IAQ Management Plan—Before Occupancy?

   A. Space Evaluation

   B. Distribution Testing

   C. Building Flush-out

   D. Air Testing

   E. Monitoring

4. Teams must provide individual comfort controls for what percentage of building occupants to earn one point for IEQc6.2 Controllability of Systems—Thermal Comfort?

   A. 50 percent

   B. 60 percent

   C. 70 percent

   D. 80 percent

   E. 90 percent

5. Upon completion of the thermal comfort survey for IEQc7.2 Thermal Comfort—Verification, when should a corrective action plan be initiated?

   A. For all problems identified

   B. For all problems identified by more than 5 percent of occupants

   C. For all problems identified by more than 10 percent of occupants

   D. For all problems identified by more than 15 percent of occupants

   E. For all problems identified by more than 20 percent of occupants

6. To earn IEQc4.1 Low-Emitting Materials—Adhesives and Sealants, with which standard does the VOC content of adhesives and sealants used on a project need to comply?

   A. SCAQMD, rule #1113

   B. Green Seal Standards

   C. MERV

   D. SCAQMD, rule #1168

   E. ASHRAE 55

7. When must the monitoring system sound an alarm in IEQc1 Outdoor Air Delivery Monitoring?

A. When airflow values or the level of CO_2 vary by 1 percent or more from the design values.

B. When airflow values or the level of CO_2 vary by 2 percent or more from the design values.

C. When airflow values or the level of CO_2 vary by 15 percent or more from the design values.

D. When airflow values or the level of CO_2 vary by 10 percent or more from the design values.

E. When airflow values or the level of CO_2 vary by 5 percent or more from the design values.

8. Which four of the following are IEQc4 Low-Emitting Material credits?

A. Paints and Coatings

B. Primers and Paints

C. Composite Wood and Agrifiber Products

D. Ceiling Systems

E. Adhesives and Sealants

F. Flooring Systems

9. Teams must provide individual comfort controls for what percentage of building occupants to earn one point for IEQc6.1 Controllability of Systems—Lighting?

A. 50 percent

B. 60 percent

C. 70 percent

D. 80 percent

E. 90 percent

10. To achieve IEQ8.1 Daylight and Views—Daylight, teams must achieve a lighting level between 25fc and 500fc for a minimum of _____ percent of regularly occupied areas.

A. 45

B. 65

C. 75

D. 85

E. 95

QUIZ ANSWERS

Lesson 6

1. **B** The referenced standard for this credit is ASHRAE 62.1—2007: Ventilation for Acceptable Indoor Air Quality. Download this standard at wwww.ashrae.org.

2. **C** CO2 monitors must be between 3' and 6' above the floor.

3. **C, D** To ensure the building is free of major contaminants, or is below recognized acceptable levels before occupancy, teams use either the building flush-out procedure or IAQ testing.

4. **A** Teams must provide individual comfort controls for 50 percent of building occupants to earn one point for IEQc6.2 Controllability of Systems—Thermal Comfort.

5. **E** A corrective action plan should be developed to address problems identified by more than 20 percent of occupants.

6. **D** To achieve IEQc4.1 Low-Emitting Materials—Adhesives and Sealants all VOC content of adhesives and sealants must comply with the South Coast Air Quality Management District, SCAQMD, rule #1168.

7. **D** The monitoring system must sound an alarm if airflow values or the level of CO2 vary by 10 percent or more from the design values.

8. **A, C, E, F** The following are Low-Emitting Material Credits: IEQc4.1 Adhesives and Sealants, IEQc4.2 Paints and Coatings, IEQc4.3 Flooring Systems and IEQc4.4 Composite Wood and Agrifiber Products.

9. **E** Team must provide individual lighting controls for 90 percent of building occupants to earn one point for IEQc6.1 Controllability of Systems—Lighting.

10. **C** To achieve IEQ8.1 Daylight and Views—Daylight, teams must achieve a lighting level between 25fc and 500fc in a minimum of 75 percent of regularly occupied areas.

INNOVATION IN DESIGN

INTRODUCTION

Innovation in Design provides teams the opportunity to earn points for using new technologies that are not addressed in the Reference Guide or for attaining exemplary performance by exceeding credit thresholds. With the growth of sustainable design, new technologies are continually entering the marketplace.

In this chapter, we will look at the credits for the **Innovation in Design** category, summarized below:

IDc1 Innovation in Design .... 1–5 Points
IDc2 LEED® Accredited Professional ... 1 Point

IDc1 INNOVATION IN DESIGN

Intent

Teams using emerging technologies or seeking exemplary performance for established credits not addressed in existing LEED® credits can have significant benefits for the environment and building occupants.

Requirements

To achieve this credit, teams can earn points through any combination of the following paths:

Path 1: Innovation in Design

One point is awarded for each innovation achieved. Up to five points will be awarded for Path 1.

Identify in writing the intent of the proposed innovation credit in addition to:

- Proposed requirements for compliance
- Proposed submittals to demonstrate compliance
- Design approach that might be used to meet the requirements

Path 2: Exemplary Performance

One point is awarded for each exemplary performance achieved with a maximum of three points awarded for Path 2. Refer to the appropriate credit descriptions for requirements. Exemplary performance credit points are available for the following credits:

- Sustainable Sites
  - SSc2 Development Density and Community Connectivity
  - SSc4 Alternative Transportation
  - SSc5 Site Development
  - SSc6 Stormwater Design
  - SSc7 Heat Island Effect
- Water Efficiency
  - WEc2 Innovative Wastewater Technologies
  - WEc3 Water Use Reduction
- Energy and Atmosphere
  - EAc1 Optimize Energy Performance
  - EAc2 On-Site Renewable Energy
  - EAc3 Enhanced Commissioning
  - EAc6 Green Power
- Materials and Resources
  - MRc2 Construction Waste Management
  - MRc3 Materials Reuse
  - MRc4 Recycled Content
  - MRc5 Regional Materials
  - MRc6 Rapidly Renewable Materials
- Indoor Environmental Quality
  - IEQc8 Daylight and Views

IDc2 LEED® ACCREDITED PROFESSIONAL

Intent

This credit is intended to encourage and support the LEED® AP's involvement in project teams. APs bring an understanding of the prerequisites and credits, their respective criteria and the coordination and documentation methods that can streamline the application and certification process.

Requirements

To achieve this credit, teams must have at least one principal participant of the project that is a LEED® AP.

Referenced Standard

The referenced standard for this credit is a Green Building Certification Institute accredited LEED® AP. The Green Building Certification Institute accreditation process can be downloaded at *www.gbci.org*.

LESSON EIGHT

8

REGIONAL PRIORITY

INTRODUCTION

Regional Priority points have been added to the LEED® 2009 rating system to encourage design teams to focus on region-specific environmental issues. USGBC Regional Councils have identified environmental zones, based on ZIP codes, and allocated six credits within each zone eligible to receive regional priority points. Information regarding the zones, broken down by state and then zip code can be downloaded at *www.usgbc.org*.

RPc1 Regional Priority .... 1–4 Points

RPc1 REGIONAL PRIORITY

Projects can earn between one and four points for addressing geographically-specific environmental priorities based on the zip code of the project. Teams can access the database of Regional Priority points and their applicability at *www.usgbc.org*.

GLOSSARY

A

Adapted plants are considered non-invasive and low-maintenance. Once established in a given habitat, they grow reliably well with only minimal winter protection, pet control fertilization, and irrigation.

Age of Air is the average amount of time that has elapsed since a sample of air molecules at a specific location has entered the building.

Air-Change Effectiveness is a measurement based on a comparison of the age of air in the occupied portions of the building, to the age of air that would exist under conditions of perfect missing of the ventilation air.

Albedo is the measure of a surface's ability to reflect sunlight. Also known as **Solar Reflectance**.

Alternative Fuel Vehicles (AFV) are vehicles that use low-polluting, non-gasoline fuels. Fuels in this group include: electricity, hydrogen, propane or compressed natural gas, liquid natural gas, methanol, and ethanol. Efficient gas-electric hybrid vehicles are also considered AFVs.

Aquatic systems are ecologically designed treatment systems that utilize biological organisms to treat wastewater.

Aquifer is an underground rock formation that readily transmits water to wells and springs.

B

Baseline is information gathered at the beginning of a design, documenting proposed conditions from which proposed variations to the design can be compared.

Basis of Design (BOD) is a document written by the design team to describe systems to be commissioned, design assumptions, applicable codes, standards, regulations, guidelines, and information necessary to achieve the OPR. This document is to be updated for each design submittal to reflect design revisions and refined details. Minimum BOD Requirements include the following:

- Primary design assumptions concerning systems to be commissioned
- Standards to be followed to achieve BOD, including codes guidelines and regulations
- Narrative descriptions for systems to be commissioned, including for HVAC&R systems, lighting, hot water systems, on-site power systems performance criteria

Best Management Practices (BMP) criteria include: a) Design in accordance with adopted local or state program standards of performance; b) Criteria compliance demonstrated through in-field performance monitoring.

Biomass typically refers to vegetation, but also refers to all of the living material in a given area.

Bioremediation is the treatment of pollutants or waste by the use of microorganisms and vegetation to break down undesirable substances in the soil and water.

Blackwater is wastewater from toilets, urinals, and kitchen sinks that contains organic materials.

Brownfield Site means real property, the expansion, redevelopment, or reuse of which may be complicated by the presence, or potential presence, of a hazardous substance, pollutant, or contaminant.

Building Density is the floor area of the building, divided by the total area of the site.

Building Footprint is the area of the site used by the building structure. Parking lot, landscaping, and non-building related structures are not included in the area considered the building footprint.

Building Interior (as it applies to Low-Emitting Materials—Painted and Coatings), refers to the inside of the weather proofing system, and is applied on site.

C

Carbon Dioxide (CO_2) is an indicator of the effectiveness of the ventilation of a space.

The **Center for Resource Solutions (CRS)** is a national not-for-profit organization that administers the Green-e Program.

Chain-of-Custody (COC) is a tracking procedure that documents the status of a product from the point of harvest, or extraction to the end user.

Chlorofluorocarbons (CFCs) are hydrocarbons that deplete the stratospheric ozone layer.

Cogeneration is the generation of electricity by capturing heat energy that otherwise would be wasted as a byproduct. It is also referred to as a combined heat and power (CHP) system.

Composite wood consists of wood or plant particles or fibers bonded together with a synthetic resin or binder.

Composting toilet is a dry plumbing fixture that contains and treats human waste via microbiological process.

The **Comprehensive Environmental Response, Compensation, and Liability Act (CERCLA)** is a tax on chemical and petroleum industries that allows federal authorities to respond to the releases of hazardous substances into the environment. Also known as **Superfund**.

Conditioned Space is the portion of the building that is heated and/or cooled for the comfort of building occupants.

Constructed Wetland is an engineered system designed to simulate natural wetland functions for water purification. Constructed wetlands are essentially treatment systems that remove contaminants from wastewaters.

Construction and Demolition (C&D) includes waste and recyclables generated from construction, renovation, and demolition, or deconstruction of pre-existing structures. Land clearing waste, such as soil, vegetation, and rocks are not included.

A **Construction IAQ Management Plan** is a document specific to a building project that outlines procedures to minimize building contamination during construction, and building flush-out prior to occupancy.

Conventional Irrigation is the most common irrigation system used in a region where the building is located. A common conventional irrigation system uses pressure to deliver water, and distributes it through sprinkler heads above ground.

Cutoff Angle is the angle formed by a line drawn from the direction of the direct light rays at the light source with respect to the vertical, beyond which no direct light is emitted.

D

Daylighting is the controlled admission of natural light into a space through glazing with the intent of reducing or eliminating electric lighting. By using solar light, daylighting creates a stimulating productive environment for building occupants.

The **Development Footprint** is an area of impact that includes building footprint, roads, parking lots, sidewalks, and outbuildings.

Drip Irrigation is a high-efficiency irrigation method where water drips to the soil from perforated tubes.

E

An **Ecosystem** is an ecological unit consisting of complex community of organisms.

Emissivity is the ratio of the radiation emitted by a surface, to the radiation emitted by a black body at the same temperature and area.

Emittance is the measure of a surface to emit radiant (thermal) energy.

Endangered Species are animal or plant species in danger of becoming extinct though all or significant portion of its range due to harmful human activities or environmental factors.

Energy Conservation Measures (ECMs) pertain to the installation of equipment or systems, or modifications of equipment or systems, for the purpose of reducing energy use and/or cost.

Energy Cost Budget (ECB) Performance Path allows flexibility by permitting the use of ECB method to evaluate building performance. The design is evaluated based on the cost of various types of energy used instead of units of energy used. An advantage of ECB is that it allows credit for innovative energy-efficient designs, such as passive solar heating and daylighting, which are not accounted for in the prescriptive or system performance paths.

ENERGY STAR® is a government/industry partnership managed by EPA and US DOE. It offers guidelines for purchasing products, and lists of Energy Star labeled items. It also provides management strategies and benchmarking software tools for buildings.

ENERGY STAR® Target Finder are tools to help architects and building owners establish aggressive, yet realistic energy targets, and rate the estimated energy use of a buildings design.

Environmental Resource Guide is a comprehensive guide documenting materials life-cycle cost analysis.

Environmental Tobacco Smoke (ETS) is also known as second-hand smoke.

Erosion is a natural process that causes the wearing away of the Earth's surface.

Evapotranspiration is the loss of water by evaporation from soil, and transpiration from plants.

Ex-Situ Remediation is the removal of contaminated soil and ground water to another location for treatment.

An **Extensive Green Roof** is a simpler green roof with a soil layer of 6 inches or less to support turf, grass, or other ground cover.

F

4-phenylcylohene (4-PCH) is an easily detected odor, commonly know as 'new carpet' odor. Emitted from materials commonly used to bind carpet fibers to its backing.

Footcandle is a measurement of illumination where one unit is equal to the light of a candle at a distance of one foot.

Forest Stewardship Council (FSC) requires observance of regional and international treaties, respects rights of indigenous people to manage their land, protects forest worker's health, and does not prohibit the use of pesticides.

Formaldehyde (CH_20) is a colorless, pungent, and irritating gas used chiefly as a disinfectant and preservative, and in synthesizing other compounds, like resins.

Full Cutoff Fixture is a lamp and fixture assembly designed with a cutoff angle of 90º so that no direct light is emitted above a horizontal plane. Technical description: Zero candela intensity at, or above, horizontal (90° above NADIR). Also, candela intensity at 80° above NADIR not to exceed a value equal to 10 percent of lamp lumens. Also Known as **Fully Shielded.**

Fully Shielded (*see* Full Cutoff Fixture)

G

Geothermal is the practice of harvesting the heat from the earth to use as an energy source.

Glare is light emitted from a luminaire with an intensity great enough to produce annoyance, discomfort, or a reduction in a viewer's ability to see.

Green-e is a program established by CRS to promote green energy products.

Greenfield is undeveloped land, or land that has not been impacted by human activity.

Green Label is the Carpet and Rug Institute's testing program to identify low VOC products.

A **Green Roof** is a rooftop planted with vegetation. **Intensive Green Roofs** have thick layers of soil (6 to 12 inches or more) that can support a broad variety of plant, or even tree species. **Extensive Green Roofs** are simpler green roofs with a soil layer of 6 inches or less to support turf, grass, or other ground cover. Also known as **Vegetative Roof.**

Green Seal is an independent, not-for-profit organization that provides product certification and purchasing guidelines for products.

GreenSpec is a Web site that lists green building products, their environmental data, manufacturer information, and provides links to additional sources.

Greywater is wastewater from lavatories, showers, bathtubs, washing machines, and sinks that are not used for disposal of hazardous or toxic ingredients, or wastes from food preparation. Greywater can be reused with only simple filtration.

Group Multi-Occupant Space includes conference rooms, classrooms, and other indoor spaces used for presentations, training, teaching, etc. Individuals in these spaces share the lighting and temperature controls.

H

Halons are a substance used in fire suppression systems and fire extinguishers in buildings. They deplete the stratospheric ozone layer.

Hardscape are components of a landscape constructed from materials other than plants, including walks, walls, and trellises made of wood, stone, or other materials.

A **Heat Island** is a developed area, or microclimate, where the temperature is higher than undeveloped areas.

A **Hybrid Vehicle** is a vehicle that uses a gasoline engine to drive an electric generator, which charges batteries that also power the vehicle.

Hydroclorofluorcarbons (HCFCs) are refrigerants that deplete the stratospheric ozone layer, but cause less damage than CFCs.

Hydrofluorocarbons (HFCs) are refrigerants that do not deplete the stratospheric ozone layer, but some HFCs do have high global warming potential.

I

IESNA RP-33 Lighting Zones

LZ1—Dark, Park, and Rural areas
LZ2—Low, Residential areas
LZ3—Medium, Commercial/Industrial, High-Density Residential areas
LZ4—High, Major City Centers, Entertainment districts

Illuminance is the amount of light falling on a surface, and can be measured in foot-candles (fc), or lux (lx).

Individual Multi-Occupant Space is typically an open-office plan. These spaces normally contain standard work stations where each individual must have comfort controls to earn Credit 6.2.

Indoor Air Quality (IAQ) is the quality of indoor air in which there are no known contaminants at harmful concentrations.

In Situ Remediation is the treatment of contaminated soil and ground water in place.

An **Intensive Green Roof** has thick layers of soil (6 to 12 inches or more) that can support a broad variety of plant, or even tree species.

Invasive Plants are one of the biggest threats to biodiversity and ecosystem stability. Non-Indigenous plant species can grow and multiply aggressively and overrun biosystems (including native/indigenous plants).

L

Landfill is a waste disposal site for the deposit of solid waste from human activities.

Light Pollution includes all adverse effects of man-made light, including sky glow, glare, light trespass, light clutter, decreased visibility at night, and wasted energy.

Light Trespass is the shining of direct light produced by a luminaire beyond the boundaries of the lot or parcel on which it is located. Obtrusive, unwanted, light due to quantity, direction, or spectrum attributes. Causes annoyance, discomfort, distraction, or loss of visibility.

Living Machine is a wastewater treatment system that uses natural bioremediation processes, such as wetlands.

Lumen is a measure of light power, also known a luminous flux, generated by a light source. One foot candle is one lumen per square foot.

Luminance is the amount of visible light leaving a point on a surface in a given direction. This "surface" can be a physical surface or an imaginary plane, and the light leaving the surface can be due to reflection, transmission, and/ or emission.

M

Metering Controls are controls that turn off water typically on faucets and showers. The devices can be either automatic or manual.

Micro-irrigation is a system designed to provide small volumes of water thru drippers, micro-jets, or small sprinklers. Drippers are installed below grade, and sprinklers and micro-jets are installed just above the ground.

Minimum Efficiency Reporting Value (MERV) measures air filtration media effectiveness. The **Montreal Protocol** is an international treaty that governs stratospheric ozone protection and research, and the production and use of ozone-depleting substances. It supplies the means of support to end production of ozone-depleting substances, including CFCs. It also provides resources to developing nations to promote the transition to ozone-safe technologies.

N

The **National Resources Defense Council (NRDC)** protects wildlife and wild places.

Native/Indigenous Plants are vegetation that grows in a region prior to settlement; they are non-invasive.

Natural Ventilation is the process of supplying and removing air without mechanical ductwork in building spaces by using openings, such as windows and doors, non-powered ventilators, and filtration processes.

Non-Occupied Space includes all rooms used by maintenance personnel that are not open for use by occupants, including janitorial, storage, equipment rooms, and closets.

Non-Regularly Occupied Space includes corridors, hallways, lobbies, break rooms, copy rooms, storage rooms, kitchens, restrooms, and stairwells.

O

Occupied Zone is the region in an occupied space from 3 inches above the floor to 72 inches above the floor, and greater than 2 feet from walls or fixed air conditioning equipment.

On-Site Wastewater Treatment is a localized treatment system that transports, stores, treats, and disposes of wastewater volumes generated on the project site.

Open Space is the property area, minus the development foot print. Open space is to be vegetated and pervious to provide habitat and other ecological benefits.

Owner Project Requirement (OPR) is a document written by the owner to outline their goals and expectations for systems to be commissioned.

P

Pervious and Permeable Surfaces both allow fluid to pass through, or are penetrable.

Plumbing Fixture Sensors are sensors that are applied to water closets, lavatories, sinks, and urinals to sense fixture use and automatically turn on and off.

A **Prescriptive Performance** approach is quick and easy to use, but can be restrictive, because it is based on worst-case scenarios.

Process Water is water that is used in chillers, boilers, cooling towers, and industrial processes.

Porous means having small holes or pours.

Postconsumer material, also known as feedstock, that is collected from businesses and residential recycling programs, as well as construction and demolition materials for reuse in another product.

Preconsumer material that is collected from the industrial process and reused for the same or similar industrial process, or is sold or traded for a different industrial process, but it is never sold as a consumer product. Does not include in-house scrap that is fed back in the same manufacturing process.

Potable Water is water that meets drinking water quality standards and is approved for human consumption by the state or local authorities having jurisdiction.

Property Area is the legal property boundary of a project, encompassing the entire site, including constructed areas and non-constructed areas. Also known as **Site Area**.

R

Recycling is the collection, reprocessing, marketing and use of materials that were diverted or recovered from the solid waste stream.

Refrigerants are fluids that absorb heat from a reservoir at low temperatures and reject heat at high temperatures. Refrigerants are the working fluids of refrigeration cycles.

Regularly Occupied Space is an area where workers are seated or standing as they work inside a building.

Relative Humidity is the ratio of partial density of water vapor in the air to the saturation density of water vapor at the same temperature.

Remediation is the process of cleaning up site contaminants.

The **Resource Conservation and Recovery Act (RCRA)** gave the EPA the authority to control hazardous wastes from cradle to grave including generation, transportation, treatment, storage, and disposal. There are some non-hazardous wastes also covered under this act.

Return Air is previously conditioned air that is removed from a space.

Reuse is a strategy to return materials to active use in the same or related capacity.

Risk Assessment is a methodology used to evaluate for potential health effects caused by contaminants in the environment. Information from the risk assessment is used to determine cleanup levels.

R-Value is the unit of measure of resistance of a substance to heat flow.

S

Salvaged or reused materials are construction materials recovered from existing building or construction sites, and reused in other buildings or construction sites. Salvaged materials include, but are not limited to, structural columns and beams, flooring, cabinetry, brick, and decorative elements.

Sedimentation is the addition of soil to bodies of water through natural and human actives. Sedimentation decreases the quality of water, and accelerates the aging process of these bodies of water.

Shielding is part of a luminaire or lamp designed to limit glare, light trespass, and/or light pollution.

Sick Building Syndrome are the effects that building occupants suffer as a direct result from their exposure to contaminated air inside a building.

Site Area (*see* Property Area)

Site Assessment is an evaluation of above-ground (including facilities) and subsurface characteristics, including the geology and hydrology of a site, to determine if a release of contaminants or pollutants has occurred, as well as the extent and concentration of the release. Information generated during a site assessment is used to support remedial action decisions.

Solar Reflectance (*see* Albedo)

Solar Reflectance Index (SRI) is a material's ability to reject solar heat.

Square Footage (sf) is the total area of a building inclusive of corridors, elevators, stairwells, and shaft spaces.

Superfund (*see* Comprehensive Environmental Response, Compensation, and Liability Act)

Supply Air is conditioned air that is delivered to a space.

Sustainable Forestry is the practice of managing forest resources to restore, enhance and sustain a full range of forest values—economic, social, and ecological.

T

The **Technology Acceptance Reciprocity Partnership (TARP)** is the accepted Protocol for in-field performance monitoring.

Tertiary Treatment of wastewater is the removal of organics, solids, and nutrients, as well as biological or chemical polishing of wastewater.

Thermal Comfort is a condition of mind experienced by building occupants expressing satisfaction with the thermal environment.

Threatened Species are animal or plant species that are likely to become endangered with the foreseeable future.

Tipping Fees are fees that are charged by a landfill for disposal of waste. Fees are typically quoted for one ton of waste.

Total Phosphorous (TP) consists of organically bound phosphates, poly-phosphates, and orthophosphates in stormwater. Most phosphorous is found in stormwater is from fertilizer.

Total Suspended Solids (TSS) are particles too small or light to be removed from stormwater by gravity and require filtration.

Trade-Off Approach allows the trade of enhanced energy efficiency in one component against decreased energy efficiency in other components. Tradeoffs typically occur within major building systems-envelope, lighting, or mechanical. The only tradeoff for mechanical systems and equipment is found in Chapter 8 of the IECC. Higher cooling equipment efficiency can be traded for a requirement for an economizer.

Transpiration is the emission of water vapor from plant leaves.

U

U-Value is a measure of the amount of heat that flows in or out of a substance when under constant conditions where there is a 1° in air temperature difference between each side of the substance. Typically used in determining the performance of a window assembly or glazing system.

The **Urban Land Institute** is an organization that promotes responsible use of land to enhance the environment.

V

Vegetative Roof (*see* Green Roof).

Ventilation is the process by which air is supplied to, and removed from, indoor space; this can be accomplished by mechanical or natural methods.

Ventilation Effectiveness refers to the movement of supply air through the occupied space.

Visible Transmittance (T_{vis}) is the ratio of total transmitted light to total incident light. In other words, it is the amount of light passing through a glazing surface, divided by the amount of light striking the glazing surface. A higher T_{vis} value indicates that a greater amount of incident light is passing though the glazing.

Volatile Organic Compounds (VOCs) are carbon compounds capable of entering the gas phase at normal room temperatures.

W

Wastewater is typically water that is discharged from kitchen sinks, showers, and bathtubs, that contains organic materials.

Waterless Urinal is a dry plumbing fixture that uses advanced hydraulic design and a buoyant fluid instead of water to maintain sanitary conditions.

Wetland Vegetation consists of plants that require saturated soils to survive, as well as certain tree and other plant species that can tolerate prolonged wet soil conditions.

X

Xeriscaping is landscaping design that is projected to minimize or negate the use of irrigation.

SAMPLE EXAMINATION

1. SSc2 Development Density and Community Connectivity considers which three of the following to be basic services?
   - **A.** Laundry
   - **B.** Gas station
   - **C.** Fitness center
   - **D.** Bus station
   - **E.** Museum
   - **F.** Car dealership

2. According to SSc3 Brownfield Redevelopment requirements, which two of the following can classify a site as a brownfield?
   - **A.** ASTM E1903—97 Phase II Environmental Site Assessment
   - **B.** ASTM Standard E408—71(1996)e1
   - **C.** CIBSE
   - **D.** ASTM E1527—05 Phase I Environmental Site Assessment
   - **E.** Local, state, or federal government agency

3. SSc7.2 Heat Island Effect—Roof requires what percentage of low-sloped roofs to have an SRI of 29 or higher?
   - **A.** 25 percent
   - **B.** 50 percent
   - **C.** 75 percent
   - **D.** 100 percent

4. To achieve SSc4.1 Alternative Transportation—Public Transportation Access, what is the required proximity for railway and bus lines to the project?
   - **A.** The project must be located within ½ mile of a railway or within ¼ mile of two bus lines.
   - **B.** The project must be located within ¼ mile of a railway or within ½ mile of two bus lines.
   - **C.** The project must be located within ½ mile of two railway lines or within ¼ mile of one bus line.
   - **D.** The project must be located within ¼ mile of two railway lines or within ½ mile of one bus line.

5. Which three of the following Sustainable Sites prerequisites and/or credits can achieve one point for exemplary performance in Innovation in Design?
   - **A.** SSp1 Construction Activity Pollution Prevention
   - **B.** SSc1 Site Selection
   - **C.** SSc2 Development Density and Community Connectivity
   - **D.** SSc3 Brownfield Redevelopment
   - **E.** SSc5 Site Development
   - **F.** SSc6 Stormwater Design

6. A manufacturing company is planning to construct a new facility on a 400-acre greenfield site they have recently purchased in Iowa. To address the company's desire to maintain as much of the existing natural habitat as possible, the project team limits site development to 10 acres and they specify site disturbance, including earthwork and clearing of vegetation, will be limited to 40 feet beyond the building perimeter, 10 feet beyond surface walkways, parking patios, and utilities that are smaller than 12 inches in diameter, 15 feet beyond primary roadway curbs, walkways, and main utility branch trenches, and 25 feet beyond constructed areas with permeable surfaces that require additional staging areas in order to limit compaction in the paved area. Which two of the following credits can the project team expect to achieve?

   A. SSc1 Site Selection

   B. SSc5.1 Site Development—Protect or Restore Habitat

   C. SSc7.1 Heat Island Effect—Nonroof

   D. SSc6.1 Stormwater Design—Quantity Control

   E. SSc5.2 Site Development—Maximize Open Space

7. What does albedo measure?

   A. A surface's ability to absorb sunlight

   B. A surface's ability to reflect sunlight

   C. A surface's ability to release sunlight

   D. A surface's ability to emit sunlight

8. What type of water is typically discarded from a lavatory that does not contain human waste or toxic substances?

   A. Potable water

   B. Brownwater

   C. Blackwater

   D. Greywater

9. What type of landscape design endeavors to eliminate the need for irrigation?

   A. Hydrology

   B. Xeriscape

   C. Drip irrigation

   D. Evapotranspiration

10. Which four of the following fixtures, fittings, and appliances are not included in the water use reduction calculation for WEp1 Water Use Reduction?

   A. Pre-rinse spray valves

   B. Commercial dishwashers

   C. Commercial clothes washers (family sized)

   D. Automatic commercial ice makers

   E. Standard and compact residential dishwashers

   F. Kitchen sink faucets

11. Which two of the following Water Efficiency prerequisites and/or credits can achieve one point for exemplary performance in Innovation in Design?

   A. WEp1 Water Use Reduction

   B. WEc1 Water Efficient Landscaping

   C. WEc2 Innovative Wastewater Technologies

   D. WEc3 Water Use Reduction

12. To calculate WEc1 Water Efficient Landscaping compliance, which four of the following pieces of information are required?

A. Landscape Coefficient

B. Microclimate Factor

C. Irrigation Factor

D. Climate Coefficient

E. Density Factor

F. Species Factor

13. Which three of the following fixtures are used in the calculation to determine compliance with WEp1 Water Use Reduction?

A. Hose bibs

B. Water closets

C. Urinals

D. Kitchen faucets

E. Drinking fountains

F. Sprinklers

14. The prerequisite EAp1 Fundamental Commissioning of Building Energy Systems requires systems that are essential to the operation of a building to be commissioned. Which four of the following energy-related systems are required to be commissioned to achieve this prerequisite?

A. HVAC&R mechanical and passive systems and their controls

B. Controls for lighting and daylighting

C. Domestic water heating systems

D. Renewable energy systems, including wind and solar

E. Passive solar systems

F. Information technology systems

G. Stormwater management systems

15. What is the referenced standard for EAc5 Measurement and Verification?

A. ANSI E–779—03

B. IPMVP, Volume III, EVO 30000.1—2006

C. EPAct 2005

D. SMACNA IAQ Guidelines, 2nd ed., Chapter 3, November 2007

16. Which two of the following pieces of information are needed to calculate the refrigerant Atmospheric Impact for EAc4 Enhanced Refrigerant Management?

A. Global warming potential

B. Life-cycle global warming potential

C. Ozone depletion potential

D. Life-cycle ozone depletion potential

E. Refrigerant leakage rate

F. End-of-life refrigerant loss

17. Which three Energy and Atmosphere prerequisites and credits cite the reference standard ASHREA 90.1—2007?

A. EAp1 Fundamental Commissioning of Building Energy Systems

B. EAp2 Minimum Energy Performance

C. EAc1 Optimized Energy Performance

D. EAc2 On-Site Renewable Energy

E. EAc5 Measurement and Verification

F. EAc6 Green Power

18. A project team is pursuing EAc2 On-Site Renewable Energy. How many points will they earn if they design a passive solar system that will constitute 5 percent of the building annual energy cost?

A. 1 point

B. 0 point

C. 3 points

D. 4 points

19. Which four of the following Energy and Atmosphere prerequisites and/or credits can achieve one point for exemplary performance in Innovation in Design?

A. EAp3 Fundamental Refrigerant Management

B. EAc1 Optimized Energy Performance

C. EAc2 On-Site Renewable Energy

D. EAc3 Enhanced Commissioning

E. EAc4 Enhanced Refrigerant Management

F. EAc5 Measurement and Verification

G. EAc6 Green Power

20. A project team is reusing an existing building, but it is determined that only certain elements of the existing building can be retained and it will not be possible to attain a credit for MRc1 Building Reuse. Will it still be possible to achieve a credit for the portion of the building they are reusing?

A. Yes, it can contribute to MRc2 Construction Waste Management.

B. Yes, it can contribute to MRc3 Materials Reuse.

C. Yes, it can contribute to MRc5 Regional Materials.

D. No.

21. What percentage of the existing building structure and envelope must be reused to earn two points for MRc1.1 Building Reuse—Maintain Existing Walls, Floor, and Roof?

A. 55 percent

B. 60 percent

C. 75 percent

D. 80 percent

E. 95 percent

22. What percentage of the existing building structure and envelope must be reused to earn three points for MRc1.1 Building Reuse—Maintain Existing Walls, Floor, and Roof?

A. 55 percent

B. 60 percent

C. 75 percent

D. 80 percent

E. 95 percent

23. Which four of the following materials may contribute to the postconsumer or preconsumer content required to achieve MRc4 Recycled Content?

A. Steel scrap

B. Reground plastic polymers

C. Clay

D. Concrete

E. Brick

F. Ceiling tile

G. Gypsum board

24. Which four of the following Materials and Resources prerequisites and/or credits can achieve one point for exemplary performance in Innovation in Design?

A. MRp1 Storage and Collection of Recyclables

B. MRc1 Building Reuse

C. MRc2 Construction Waste Management

D. MRc3 Materials Reuse

E. MRc4 Recycled Content

F. MRc5 Regional Materials

25. A project team is considering pursing MRc6 Rapidly Renewable Materials for their project. Which four of the following materials will contribute to the credit requirements?

A. Bamboo

B. Cork

C. Natural rubber

D. Cotton

E. Cedar

F. Steel

26. What is the minimum required MERV value for air filtration media used in permanently installed air handlers utilized during the construction and pre-occupancy phases to achieve IEQc3.1 Construction IAQ Management Plan—During Construction?

A. 18

B. 9

C. 8

D. 13

E. 3

27. If the building is mechanically ventilated what is the minimum required MERV value for air filtration media installed prior to occupancy to achieve IEQc5 Indoor Chemical and Pollutant Source Control?

A. 6

B. 13

C. 8

D. 18

E. 9

28. A sporting goods company has hired an architectural firm to design their new headquarters in northern Wisconsin. To address the client's concern for building occupant health and well-being, the design team has focused on strategies to maintain a high indoor environmental quality. They design the building to isolate spaces where hazardous gases and chemicals are located, include operable windows for all easily reached windows in personal and open offices, and increased outdoor air ventilation rates by 35 percent above ASHRAE Standard 62.1—2007. Which three of the following credits can benefit from the strategies described above?

A. IEQc1

B. IEQc2

C. IEQc3.2

D. IEQc5

E. IEQc6.1

F. IEQc6.2

29. What are the four environmental factors that are addressed in ASHRAE Standard 55—2004, Thermal Environmental Conditions for Human Occupancy?

A. Activity

B. Temperature

C. Clothing

D. Thermal radiation

E. Humidity

F. Motion

G. Air speed

30. IEQc4.3 Low-Emitting Materials refer to which one of the following credits?

A. Paints and Coatings

B. Composite Wood and Agrifiber Products

C. Adhesives and Sealants

D. Flooring Systems

31. Which three of the following options are valid approaches to determine if a project can qualify for IEQc8.1 Daylight and Views—Daylight?

A. Simulation
B. Measurement
C. Glazing factor
D. Prescriptive
E. Plan view and section view analysis

32. ASHRAE 90.1—2007 applies to which four of the following prerequisites and/or credits?

A. IEQc6.1 Controllability of Systems—Lighting
B. EAc2 On-site Renewable Energy
C. EAc6 Green Power
D. EAp2 Minimum Energy Performance
E. SSc8 Light Pollution Reduction
F. EAc1 Optimized Energy Performance
G. EAp1 Fundamental Commissioning of Building Energy Systems

33. ASHRAE 52–2—1999 referenced standard applies to which two of the following prerequisites and/or credits?

A. IEQc5 Indoor Chemical Pollutant Source Control
B. IEQc3.1 Construction IAQ Management Plan—During Construction
C. IEQc3.2 Construction IAQ Management Plan—Before Occupancy
D. IEQc1 Outdoor Air Delivery Monitoring
E. IEQc6.2 Controllability of System—Thermal Comfort

34. Which of the following can help teams reduce heat island effect?

A. Emissivity
B. Evapotranspiration
C. Eutrophication
D. Increased albedo
E. Exfiltration
F. Deciduous trees

35. What is the primary focus for ASHRAE Standard 55—2004?

A. Building energy standards
B. Building ventilation filters/particulate size
C. Thermal comfort
D. Minimum ventilation
E. Air change effectiveness

36. Which four of the following spaces are not considered regularly occupied spaces for achieving IEQc8.2 Daylight and Views—Views?

A. Copy rooms
B. Gymnasiums
C. Restrooms
D. Kitchens
E. Mechanical rooms

37. IEQc5 Indoor Chemical and Pollutant Source Control requires which three of the following spaces to be isolated and maintained at negative pressure from other occupied spaces in the building?

A. Restrooms

B. Copy rooms

C. Garages

D. Laundry rooms

E. Smoking rooms

F. Private offices

38. Which three of the following Indoor Environmental Quality prerequisites and/or credits can achieve one point for exemplary performance in Innovation in Design?

A. IEQp1 Minimum Indoor Air Quality Performance

B. IEQc2 Increased Ventilation

C. IEQc4 Low—Emitting Materials

D. IEQc5 Indoor Chemical Pollutant Source Control

E. IEQc8 Daylight and Views

EXAMINATION ANSWERS

1. **A, C, E** For a complete list of basic services, see SSc2 Development Density and Community Connectivity.

2. **A, E** To achieve this credit, teams need to develop a site that is contaminated or classified as a brownfield as defined by ASTM E1903-97 Phase II Environmental Site Assessment or identified as a brownfield by a local, state, or federal government agency.

3. **C** Teams using reflective roof materials can achieve SSc7.2 Heat Island Effect—Roof by designing 75 percent of roof surfaces to be high-albedo. Low-sloped roofs must have an SRI of 78 and steep-sloped roofs must have an SRI of 29.

4. **A** To attain SSc4.1 Alternative Transportation—Public Transportation Access, the project must be located within ½ mile of a railway or within ¼ mile of two bus lines, as a pedestrian would walk.

5. **C, E, F** The following Sustainable Sites credits can achieve one point for exemplary performance in Innovation in Design:

   - SSc2 Development Density and Community Connectivity
   - SSc4 Alternative Transportation
   - Sc5 Site Development
   - SSc6 Stormwater Design
   - SSc7 Heat Island Effect

6. **B, E** The team's approach will fulfill the requirements of SSc5.1 Site Development—Protect or Restore Habitat by limiting greenfield site disturbance, including earthwork and clearing of vegetation, to 40 feet beyond the building perimeter, 10 feet beyond surface walkways, parking patios, and utilities that are smaller than 12 inches in diameter, 15 feet beyond primary roadway curbs, walkways, and main utility branch trenches, and 25 feet beyond constructed areas with permeable surfaces that require additional staging areas in order to limit compaction in the paved area. The requirements of SSc5.2 Site Development—Maximize Open Space are addressed by the team by designating open space area adjacent to the building that is equal to the development footprint. In this case, the open space is considerably larger than the development footprint.

7. **B** Albedo is the measure of a surface's ability to reflect sunlight.

8. **D** Greywater is wastewater from lavatories, showers, bathtubs, washing machines, and sinks that are not used for disposal of hazardous or toxic ingredients or wastes from food preparation. Greywater can be reused with only simple filtration.

9. **B** Xeriscaping is landscaping design that is projected to minimize or negate the use of irrigation.

10. **B, C, D, E** The following fixtures, fittings, and appliances are not included in the water use reduction calculation for WEp1 Water Use Reduction:

    - Commercial steam cookers
    - Commercial dishwashers
    - Commercial clothes washers (family sized)
    - Automatic commercial ice makers
    - Residential clothes washers
    - Standard and compact residential dishwashers

11. C, D The following Water Efficiency credits can achieve one point for exemplary performance in Innovation in Design:

- WEc2 Innovative Wastewater Technologies
- WEc3 Water Use Reduction

12. A, B, E, F The calculation to determine credit compliance may not be asked on the exam, but it is important to know that species, density and microclimate factors, and the landscape coefficient are required for the calculation.

13. B, C, D Hose bibs, sprinklers, and drinking fountains are not included in the calculation for WEp1 Water Use Reduction.

14. A, B, C, D EAp1 Fundamental Commissioning of Building Energy Systems requires the following energy-related systems to be commissioned:

- HVAC&R mechanical and passive systems and their controls
- Controls for lighting and daylighting
- Domestic water heating systems
- Renewable energy systems, including wind and solar

15. B The referenced standard for EAc5 Measurement and Verification is the International Performance Measurement and Verification Protocol, Volume III, EVO 30000.1—2006, Concepts and Options for Determining Energy Saving in New Construction, effective January 2006.

16. B, D Life-cycle global warming potential and life-cycle ozone depletion potential are needed to calculate the refrigerant atmospheric impact.

Refrigerant Atmospheric Impact =

$$\sum LCGWP + LCODP \times 10^5 \leq 100$$

17. B, C, D ASHRAE 90.1—2007 establishes the minimum requirements for energy efficient design in buildings. Most prerequisites and credits have referenced standards; each is specific to the prerequisites or credits requirements.

18. B EAc2 On-Site Renewable Energy does not recognize passive solar systems as an energy source.

19. B, C, D, G The following Energy and Atmosphere credits can achieve one point for exemplary performance in Innovation in Design:

- EAc1 Optimized Energy Performance
- EAc2 On-Site Renewable Energy
- EAc3 Enhanced Commissioning
- EAc6 Green Power

20. A If the salvaged components from the existing building will not meet the requirements for MRc1 Building Reuse, the materials can still contribute toward MRc2 Construction Waste Management.

21. C Teams must reuse 75 percent of the existing building structure and envelope to earn two points for MRc1.1 Building Reuse—Maintain Existing Walls, Floor and Roof.

22. E Teams pursuing MRc1.1 Building Reuse-Maintain Existing Walls, Floors and Roof can earn up to three points for maintaining existing building components as follows: one point for achieving 55 percent, two points for achieving 75 percent and three points for achieving 95 percent.

23. D, E, F, G Many common construction materials such as steel, brick, gypsum board and acoustical ceiling tile contain recycled content due to their manufacturing processes. Materials that are reclaimed from the same manufacturing processes do not contribute to this credit because, when waste is incorporated back into the same manufacturing process, there is no material diverted from the waste stream.

24. C, D, E, F The following Materials and Resources credits can achieve one point for exemplary performance in Innovation in Design:

- MRc2 Construction Waste Management
- MRc3 Materials Reuse
- MRc4 Recycled Content
- MRc5 Regional Materials
- MRc6 Rapidly Renewable Materials

25. A, B, C, D Rapidly renewable materials include bamboo, cotton, natural rubber and cork. Wood, steel, stone and minerals cannot be planted and harvested in a cycle of ten years or less.

26. C IEQc3.1 Construction IAQ Management Plan—During Construction requires filtration media with minimum efficiency MERV of 8 to be installed at each return air grille for all permanently installed air handlers operated during construction.

27. B IEQc5 Indoor Chemical and Pollutant Source Control, if the building is mechanically ventilated, replace air filtration media for both return and outside air that is delivered as supply air with filters having a MERV of at least 13 prior to occupancy.

28. B, D, F IEQc2 Increased Ventilation requires a minimum 30 percent increase over ASHRAE Standard 62.1—2007, IEQc5 Indoor Chemical and Pollutant Source Control requires isolation of spaces where hazardous gases and chemicals are located, and IEQc6.2 Controllability of Systems—Thermal Comfort requires comfort controls for 50 percent of building occupants which includes operable windows.

29. B, D, E, G ASHRAE Standard 55—2004 specifies environmental and human factors. The environmental factors are temperature, thermal radiation, humidity, and air speed. The human factors are activity and clothing.

30. D IEQc4.3 is Low-Emitting Materials—Flooring Systems.

31. A, B, D There are four options available to determine if the requirements for IEQc8.1 Daylight and Views—Daylight are met. They are: (1) simulation, (2) prescriptive, (3) measurement, and (4) combination of any of the preceding.

32. B, D, E, F ASHRAE 90.1—2007 establishes the minimum requirements for energy efficient design in buildings. Most prerequisites and credits have referenced standards; each is specific to the prerequisites or credits requirements.

33. A, B ASHRAE 52.2—1999 provides methods for testing and determination of MERV ratings for filtration media. Most prerequisites and credits have referenced standards; each is specific to the prerequisites or credits requirements.

34. B, D, F Heat island effect can be reduced by providing shading, landscaping, and hardscape with a high-albedo.

35. C The primary focus of ASHRAE Standard 55—2004, Thermal Comfort Conditions for Human Occupancy, is human thermal comfort.

36. A, C, E The requirements for IEQc8.2 Daylight and Views—Views, do not consider copy, storage, mechanical, laundry, and toilet rooms to be regularly occupied areas.

37. B, C, D IEQc5 Indoor Chemical and Pollutant Source Control requires copying, printing, garages, and laundry spaces to be isolated and maintained at negative pressure from other occupied spaces in the building.

38. E The following Indoor Environmental Quality credit can achieve one point for exemplary performance in Innovation in Design: IEQc8 Daylight and Views.

INDEX

P

R

S

T

U

V

W

Notes

Notes

Notes

Notes

Notes

Notes

Notes